Biodiversity Conservation: A Very Short Introduction

VERY SHORT INTRODUCTIONS are for anyone wanting a stimulating and accessible way into a new subject. They are written by experts, and have been translated into more than 45 different languages.

The series began in 1995, and now covers a wide variety of topics in every discipline. The VSI library currently contains over 700 volumes—a Very Short Introduction to everything from Psychology and Philosophy of Science to American History and Relativity—and continues to grow in every subject area.

Very Short Introductions available now:

ABOLITIONISM Richard S. Newman
THE ABRAHAMIC RELIGIONS Charles L. Cohen
ACCOUNTING Christopher Nobes
ADDICTION Keith Humphreys
ADOLESCENCE Peter K. Smith
THEODOR W. ADORNO Andrew Bowie
ADVERTISING Winston Fletcher
AERIAL WARFARE Frank Ledwidge
AESTHETICS Bence Nanay
AFRICAN AMERICAN HISTORY Jonathan Scott Holloway
AFRICAN AMERICAN RELIGION Eddie S. Glaude Jr
AFRICAN HISTORY John Parker and Richard Rathbone
AFRICAN POLITICS Ian Taylor
AFRICAN RELIGIONS Jacob K. Olupona
AGEING Nancy A. Pachana
AGNOSTICISM Robin Le Poidevin
AGRICULTURE Paul Brassley and Richard Soffe
ALEXANDER THE GREAT Hugh Bowden
ALGEBRA Peter M. Higgins
AMERICAN BUSINESS HISTORY Walter A. Friedman
AMERICAN CULTURAL HISTORY Eric Avila
AMERICAN FOREIGN RELATIONS Andrew Preston
AMERICAN HISTORY Paul S. Boyer
AMERICAN IMMIGRATION David A. Gerber
AMERICAN INTELLECTUAL HISTORY Jennifer Ratner-Rosenhagen
THE AMERICAN JUDICIAL SYSTEM Charles L. Zelden
AMERICAN LEGAL HISTORY G. Edward White
AMERICAN MILITARY HISTORY Joseph T. Glatthaar
AMERICAN NAVAL HISTORY Craig L. Symonds
AMERICAN POETRY David Caplan
AMERICAN POLITICAL HISTORY Donald Critchlow
AMERICAN POLITICAL PARTIES AND ELECTIONS L. Sandy Maisel
AMERICAN POLITICS Richard M. Valelly
THE AMERICAN PRESIDENCY Charles O. Jones
THE AMERICAN REVOLUTION Robert J. Allison
AMERICAN SLAVERY Heather Andrea Williams
THE AMERICAN SOUTH Charles Reagan Wilson
THE AMERICAN WEST Stephen Aron
AMERICAN WOMEN'S HISTORY Susan Ware
AMPHIBIANS T. S. Kemp
ANAESTHESIA Aidan O'Donnell

ANALYTIC PHILOSOPHY Michael Beaney
ANARCHISM Alex Prichard
ANCIENT ASSYRIA Karen Radner
ANCIENT EGYPT Ian Shaw
ANCIENT EGYPTIAN ART AND ARCHITECTURE Christina Riggs
ANCIENT GREECE Paul Cartledge
ANCIENT GREEK AND ROMAN SCIENCE Liba Taub
THE ANCIENT NEAR EAST Amanda H. Podany
ANCIENT PHILOSOPHY Julia Annas
ANCIENT WARFARE Harry Sidebottom
ANGELS David Albert Jones
ANGLICANISM Mark Chapman
THE ANGLO-SAXON AGE John Blair
ANIMAL BEHAVIOUR Tristram D. Wyatt
THE ANIMAL KINGDOM Peter Holland
ANIMAL RIGHTS David DeGrazia
ANSELM Thomas Williams
THE ANTARCTIC Klaus Dodds
ANTHROPOCENE Erle C. Ellis
ANTISEMITISM Steven Beller
ANXIETY Daniel Freeman and Jason Freeman
THE APOCRYPHAL GOSPELS Paul Foster
APPLIED MATHEMATICS Alain Goriely
THOMAS AQUINAS Fergus Kerr
ARBITRATION Thomas Schultz and Thomas Grant
ARCHAEOLOGY Paul Bahn
ARCHITECTURE Andrew Ballantyne
THE ARCTIC Klaus Dodds and Jamie Woodward
HANNAH ARENDT Dana Villa
ARISTOCRACY William Doyle
ARISTOTLE Jonathan Barnes
ART HISTORY Dana Arnold
ART THEORY Cynthia Freeland
ARTIFICIAL INTELLIGENCE Margaret A. Boden
ASIAN AMERICAN HISTORY Madeline Y. Hsu
ASTROBIOLOGY David C. Catling
ASTROPHYSICS James Binney
ATHEISM Julian Baggini
THE ATMOSPHERE Paul I. Palmer
AUGUSTINE Henry Chadwick
JANE AUSTEN Tom Keymer
AUSTRALIA Kenneth Morgan
AUTISM Uta Frith
AUTOBIOGRAPHY Laura Marcus
THE AVANT GARDE David Cottington
THE AZTECS Davíd Carrasco
BABYLONIA Trevor Bryce
BACTERIA Sebastian G. B. Amyes
BANKING John Goddard and John O. S. Wilson
BARTHES Jonathan Culler
THE BEATS David Sterritt
BEAUTY Roger Scruton
LUDWIG VAN BEETHOVEN Mark Evan Bonds
BEHAVIOURAL ECONOMICS Michelle Baddeley
BESTSELLERS John Sutherland
THE BIBLE John Riches
BIBLICAL ARCHAEOLOGY Eric H. Cline
BIG DATA Dawn E. Holmes
BIOCHEMISTRY Mark Lorch
BIODIVERSITY CONSERVATION David W. Macdonald
BIOGEOGRAPHY Mark V. Lomolino
BIOGRAPHY Hermione Lee
BIOMETRICS Michael Fairhurst
ELIZABETH BISHOP Jonathan F. S. Post
BLACK HOLES Katherine Blundell
BLASPHEMY Yvonne Sherwood
BLOOD Chris Cooper
THE BLUES Elijah Wald
THE BODY Chris Shilling
THE BOHEMIANS David Weir
NIELS BOHR J. L. Heilbron
THE BOOK OF COMMON PRAYER Brian Cummings
THE BOOK OF MORMON Terryl Givens
BORDERS Alexander C. Diener and Joshua Hagen
THE BRAIN Michael O'Shea
BRANDING Robert Jones
THE BRICS Andrew F. Cooper

BRITISH CINEMA Charles Barr
THE BRITISH CONSTITUTION Martin Loughlin
THE BRITISH EMPIRE Ashley Jackson
BRITISH POLITICS Tony Wright
BUDDHA Michael Carrithers
BUDDHISM Damien Keown
BUDDHIST ETHICS Damien Keown
BYZANTIUM Peter Sarris
CALVINISM Jon Balserak
ALBERT CAMUS Oliver Gloag
CANADA Donald Wright
CANCER Nicholas James
CAPITALISM James Fulcher
CATHOLICISM Gerald O'Collins
CAUSATION Stephen Mumford and Rani Lill Anjum
THE CELL Terence Allen and Graham Cowling
THE CELTS BarryCunliffe
CHAOS Leonard Smith
GEOFFREY CHAUCER David Wallace
CHEMISTRY Peter Atkins
CHILD PSYCHOLOGY Usha Goswami
CHILDREN'S LITERATURE Kimberley Reynolds
CHINESE LITERATURE Sabina Knight
CHOICE THEORY Michael Allingham
CHRISTIAN ART Beth Williamson
CHRISTIAN ETHICS D. Stephen Long
CHRISTIANITY Linda Woodhead
CIRCADIAN RHYTHMS Russell Foster and Leon Kreitzman
CITIZENSHIP Richard Bellamy
CITY PLANNING Carl Abbott
CIVIL ENGINEERING David Muir Wood
THE CIVIL RIGHTS MOVEMENT Thomas C. Holt
CLASSICAL LITERATURE William Allan
CLASSICAL MYTHOLOGY Helen Morales
CLASSICS Mary Beard and John Henderson
CLAUSEWITZ Michael Howard
CLIMATE Mark Maslin
CLIMATE CHANGE Mark Maslin
CLINICAL PSYCHOLOGY Susan Llewelyn and Katie Aafjes-van Doorn
COGNITIVE BEHAVIOURAL THERAPY Freda McManus
COGNITIVE NEUROSCIENCE Richard Passingham
THE COLD WAR Robert J. McMahon
COLONIAL AMERICA Alan Taylor
COLONIAL LATIN AMERICAN LITERATURE Rolena Adorno
COMBINATORICS Robin Wilson
COMEDY Matthew Bevis
COMMUNISM Leslie Holmes
COMPARATIVE LITERATURE Ben Hutchinson
COMPETITION AND ANTITRUST LAW Ariel Ezrachi
COMPLEXITY John H. Holland
THE COMPUTER Darrel Ince
COMPUTER SCIENCE Subrata Dasgupta
CONCENTRATION CAMPS Dan Stone
CONDENSED MATTER PHYSICS Ross H. McKenzie
CONFUCIANISM Daniel K. Gardner
THE CONQUISTADORS Matthew Restall and Felipe Fernández-Armesto
CONSCIENCE Paul Strohm
CONSCIOUSNESS Susan Blackmore
CONTEMPORARY ART Julian Stallabrass
CONTEMPORARY FICTION Robert Eaglestone
CONTINENTAL PHILOSOPHY Simon Critchley
COPERNICUS Owen Gingerich
CORAL REEFS Charles Sheppard
CORPORATE SOCIAL RESPONSIBILITY Jeremy Moon
CORRUPTION Leslie Holmes
COSMOLOGY Peter Coles
COUNTRY MUSIC Richard Carlin
CREATIVITY Vlad Glăveanu
CRIME FICTION Richard Bradford
CRIMINAL JUSTICE Julian V. Roberts
CRIMINOLOGY Tim Newburn
CRITICAL THEORY Stephen Eric Bronner
THE CRUSADES Christopher Tyerman

CRYPTOGRAPHY Fred Piper and Sean Murphy
CRYSTALLOGRAPHY A. M. Glazer
THE CULTURAL REVOLUTION Richard Curt Kraus
DADA AND SURREALISM David Hopkins
DANTE Peter Hainsworth and David Robey
DARWIN Jonathan Howard
THE DEAD SEA SCROLLS Timothy H. Lim
DECADENCE David Weir
DECOLONIZATION Dane Kennedy
DEMENTIA Kathleen Taylor
DEMOCRACY Naomi Zack
DEMOGRAPHY Sarah Harper
DEPRESSION Jan Scott and Mary Jane Tacchi
DERRIDA Simon Glendinning
DESCARTES Tom Sorell
DESERTS Nick Middleton
DESIGN John Heskett
DEVELOPMENT Ian Goldin
DEVELOPMENTAL BIOLOGY Lewis Wolpert
THE DEVIL Darren Oldridge
DIASPORA Kevin Kenny
CHARLES DICKENS Jenny Hartley
DICTIONARIES Lynda Mugglestone
DINOSAURS David Norman
DIPLOMATIC HISTORY Joseph M. Siracusa
DOCUMENTARY FILM Patricia Aufderheide
DREAMING J. Allan Hobson
DRUGS Les Iversen
DRUIDS Barry Cunliffe
DYNASTY Jeroen Duindam
DYSLEXIA Margaret J. Snowling
EARLY MUSIC Thomas Forrest Kelly
THE EARTH Martin Redfern
EARTH SYSTEM SCIENCE Tim Lenton
ECOLOGY Jaboury Ghazoul
ECONOMICS Partha Dasgupta
EDUCATION Gary Thomas
EGYPTIAN MYTH Geraldine Pinch
EIGHTEENTH-CENTURY BRITAIN Paul Langford
THE ELEMENTS Philip Ball
EMOTION Dylan Evans
EMPIRE Stephen Howe
EMPLOYMENT LAW David Cabrelli
ENERGY SYSTEMS Nick Jenkins
ENGELS Terrell Carver
ENGINEERING David Blockley
THE ENGLISH LANGUAGE Simon Horobin
ENGLISH LITERATURE Jonathan Bate
THE ENLIGHTENMENT John Robertson
ENTREPRENEURSHIP Paul Westhead and Mike Wright
ENVIRONMENTAL ECONOMICS Stephen Smith
ENVIRONMENTAL ETHICS Robin Attfield
ENVIRONMENTAL LAW Elizabeth Fisher
ENVIRONMENTAL POLITICS Andrew Dobson
ENZYMES Paul Engel
EPICUREANISM Catherine Wilson
EPIDEMIOLOGY Rodolfo Saracci
ETHICS Simon Blackburn
ETHNOMUSICOLOGY Timothy Rice
THE ETRUSCANS Christopher Smith
EUGENICS Philippa Levine
THE EUROPEAN UNION Simon Usherwood and John Pinder
EUROPEAN UNION LAW Anthony Arnull
EVANGELICALISM John G. Stackhouse Jr.
EVIL Luke Russell
EVOLUTION Brian and Deborah Charlesworth
EXISTENTIALISM Thomas Flynn
EXPLORATION Stewart A. Weaver
EXTINCTION Paul B. Wignall
THE EYE Michael Land
FAIRY TALE Marina Warner
FAMILY LAW Jonathan Herring
MICHAEL FARADAY Frank A. J. L. James
FASCISM Kevin Passmore
FASHION Rebecca Arnold
FEDERALISM Mark J. Rozell and Clyde Wilcox
FEMINISM Margaret Walters

FILM Michael Wood
FILM MUSIC Kathryn Kalinak
FILM NOIR James Naremore
FIRE Andrew C. Scott
THE FIRST WORLD WAR
Michael Howard
FLUID MECHANICS Eric Lauga
FOLK MUSIC Mark Slobin
FOOD John Krebs
FORENSIC PSYCHOLOGY
David Canter
FORENSIC SCIENCE Jim Fraser
FORESTS Jaboury Ghazoul
FOSSILS Keith Thomson
FOUCAULT Gary Gutting
THE FOUNDING FATHERS
R. B. Bernstein
FRACTALS Kenneth Falconer
FREE SPEECH Nigel Warburton
FREE WILL Thomas Pink
FREEMASONRY Andreas Önnerfors
FRENCH LITERATURE John D. Lyons
FRENCH PHILOSOPHY
Stephen Gaukroger and Knox Peden
THE FRENCH REVOLUTION
William Doyle
FREUD Anthony Storr
FUNDAMENTALISM Malise Ruthven
FUNGI Nicholas P. Money
THE FUTURE Jennifer M. Gidley
GALAXIES John Gribbin
GALILEO Stillman Drake
GAME THEORY Ken Binmore
GANDHI Bhikhu Parekh
GARDEN HISTORY Gordon Campbell
GENES Jonathan Slack
GENIUS Andrew Robinson
GENOMICS John Archibald
GEOGRAPHY John Matthews and
David Herbert
GEOLOGY Jan Zalasiewicz
GEOMETRY Maciej Dunajski
GEOPHYSICS William Lowrie
GEOPOLITICS Klaus Dodds
GERMAN LITERATURE Nicholas Boyle
GERMAN PHILOSOPHY
Andrew Bowie
THE GHETTO Bryan Cheyette
GLACIATION David J. A. Evans
GLOBAL CATASTROPHES Bill McGuire
GLOBAL ECONOMIC HISTORY
Robert C. Allen
GLOBAL ISLAM Nile Green
GLOBALIZATION Manfred B. Steger
GOD John Bowker
GÖDEL'S THEOREM A. W. Moore
GOETHE Ritchie Robertson
THE GOTHIC Nick Groom
GOVERNANCE Mark Bevir
GRAVITY Timothy Clifton
THE GREAT DEPRESSION AND
THE NEW DEAL Eric Rauchway
HABEAS CORPUS Amanda L. Tyler
HABERMAS James Gordon Finlayson
THE HABSBURG EMPIRE
Martyn Rady
HAPPINESS Daniel M. Haybron
THE HARLEM RENAISSANCE
Cheryl A. Wall
THE HEBREW BIBLE AS
LITERATURE Tod Linafelt
HEGEL Peter Singer
HEIDEGGER Michael Inwood
THE HELLENISTIC AGE
Peter Thonemann
HEREDITY John Waller
HERMENEUTICS Jens Zimmermann
HERODOTUS Jennifer T. Roberts
HIEROGLYPHS Penelope Wilson
HINDUISM Kim Knott
HISTORY John H. Arnold
THE HISTORY OF ASTRONOMY
Michael Hoskin
THE HISTORY OF CHEMISTRY
William H. Brock
THE HISTORY OF CHILDHOOD
James Marten
THE HISTORY OF CINEMA
Geoffrey Nowell-Smith
THE HISTORY OF COMPUTING
Doron Swade
THE HISTORY OF EMOTIONS
Thomas Dixon
THE HISTORY OF LIFE Michael Benton
THE HISTORY OF MATHEMATICS
Jacqueline Stedall
THE HISTORY OF MEDICINE
William Bynum
THE HISTORY OF PHYSICS
J. L. Heilbron

THE HISTORY OF POLITICAL THOUGHT Richard Whatmore
THE HISTORY OF TIME Leofranc Holford-Strevens
HIV AND AIDS Alan Whiteside
HOBBES Richard Tuck
HOLLYWOOD Peter Decherney
THE HOLY ROMAN EMPIRE Joachim Whaley
HOME Michael Allen Fox
HOMER Barbara Graziosi
HORMONES Martin Luck
HORROR Darryl Jones
HUMAN ANATOMY Leslie Klenerman
HUMAN EVOLUTION Bernard Wood
HUMAN PHYSIOLOGY Jamie A. Davies
HUMAN RESOURCE MANAGEMENT Adrian Wilkinson
HUMAN RIGHTS Andrew Clapham
HUMANISM Stephen Law
HUME James A. Harris
HUMOUR Noël Carroll
IBN SĪNĀ (AVICENNA) Peter Adamson
THE ICE AGE Jamie Woodward
IDENTITY Florian Coulmas
IDEOLOGY Michael Freeden
THE IMMUNE SYSTEM Paul Klenerman
INDIAN CINEMA Ashish Rajadhyaksha
INDIAN PHILOSOPHY Sue Hamilton
THE INDUSTRIAL REVOLUTION Robert C. Allen
INFECTIOUS DISEASE Marta L. Wayne and Benjamin M. Bolker
INFINITY Ian Stewart
INFORMATION Luciano Floridi
INNOVATION Mark Dodgson and David Gann
INTELLECTUAL PROPERTY Siva Vaidhyanathan
INTELLIGENCE Ian J. Deary
INTERNATIONAL LAW Vaughan Lowe
INTERNATIONAL MIGRATION Khalid Koser
INTERNATIONAL RELATIONS Christian Reus-Smit
INTERNATIONAL SECURITY Christopher S. Browning
INSECTS Simon Leather
IRAN Ali M. Ansari
ISLAM Malise Ruthven
ISLAMIC HISTORY Adam Silverstein
ISLAMIC LAW Mashood A. Baderin
ISOTOPES Rob Ellam
ITALIAN LITERATURE Peter Hainsworth and David Robey
HENRY JAMES Susan L. Mizruchi
JAPANESE LITERATURE Alan Tansman
JESUS Richard Bauckham
JEWISH HISTORY David N. Myers
JEWISH LITERATURE Ilan Stavans
JOURNALISM Ian Hargreaves
JAMES JOYCE Colin MacCabe
JUDAISM Norman Solomon
JUNG Anthony Stevens
THE JURY Renée Lettow Lerner
KABBALAH Joseph Dan
KAFKA Ritchie Robertson
KANT Roger Scruton
KEYNES Robert Skidelsky
KIERKEGAARD Patrick Gardiner
KNOWLEDGE Jennifer Nagel
THE KORAN Michael Cook
KOREA Michael J. Seth
LAKES Warwick F. Vincent
LANDSCAPE ARCHITECTURE Ian H. Thompson
LANDSCAPES AND GEOMORPHOLOGY Andrew Goudie and Heather Viles
LANGUAGES Stephen R. Anderson
LATE ANTIQUITY Gillian Clark
LAW Raymond Wacks
THE LAWS OF THERMODYNAMICS Peter Atkins
LEADERSHIP Keith Grint
LEARNING Mark Haselgrove
LEIBNIZ Maria Rosa Antognazza
C. S. LEWIS James Como
LIBERALISM Michael Freeden
LIGHT Ian Walmsley
LINCOLN Allen C. Guelzo
LINGUISTICS Peter Matthews
LITERARY THEORY Jonathan Culler
LOCKE John Dunn
LOGIC Graham Priest

LOVE Ronald de Sousa
MARTIN LUTHER Scott H. Hendrix
MACHIAVELLI Quentin Skinner
MADNESS Andrew Scull
MAGIC Owen Davies
MAGNA CARTA Nicholas Vincent
MAGNETISM Stephen Blundell
MALTHUS Donald Winch
MAMMALS T. S. Kemp
MANAGEMENT John Hendry
NELSON MANDELA Elleke Boehmer
MAO Delia Davin
MARINE BIOLOGY Philip V. Mladenov
MARKETING Kenneth Le Meunier-FitzHugh
THE MARQUIS DE SADE John Phillips
MARTYRDOM Jolyon Mitchell
MARX Peter Singer
MATERIALS Christopher Hall
MATHEMATICAL ANALYSIS Richard Earl
MATHEMATICAL FINANCE Mark H. A. Davis
MATHEMATICS Timothy Gowers
MATTER Geoff Cottrell
THE MAYA Matthew Restall and Amara Solari
THE MEANING OF LIFE Terry Eagleton
MEASUREMENT David Hand
MEDICAL ETHICS Michael Dunn and Tony Hope
MEDICAL LAW Charles Foster
MEDIEVAL BRITAIN John Gillingham and Ralph A. Griffiths
MEDIEVAL LITERATURE Elaine Treharne
MEDIEVAL PHILOSOPHY John Marenbon
MEMORY Jonathan K. Foster
METAPHYSICS Stephen Mumford
METHODISM William J. Abraham
THE MEXICAN REVOLUTION Alan Knight
MICROBIOLOGY Nicholas P. Money
MICROBIOMES Angela E. Douglas
MICROECONOMICS Avinash Dixit
MICROSCOPY Terence Allen
THE MIDDLE AGES Miri Rubin
MILITARY JUSTICE Eugene R. Fidell
MILITARY STRATEGY Antulio J. Echevarria II
JOHN STUART MILL Gregory Claeys
MINERALS David Vaughan
MIRACLES Yujin Nagasawa
MODERN ARCHITECTURE Adam Sharr
MODERN ART David Cottington
MODERN BRAZIL Anthony W. Pereira
MODERN CHINA Rana Mitter
MODERN DRAMA Kirsten E. Shepherd-Barr
MODERN FRANCE Vanessa R. Schwartz
MODERN INDIA Craig Jeffrey
MODERN IRELAND Senia Pašeta
MODERN ITALY Anna Cento Bull
MODERN JAPAN Christopher Goto-Jones
MODERN LATIN AMERICAN LITERATURE Roberto González Echevarría
MODERN WAR Richard English
MODERNISM Christopher Butler
MOLECULAR BIOLOGY Aysha Divan and Janice A. Royds
MOLECULES Philip Ball
MONASTICISM Stephen J. Davis
THE MONGOLS Morris Rossabi
MONTAIGNE William M. Hamlin
MOONS David A. Rothery
MORMONISM Richard Lyman Bushman
MOUNTAINS Martin F. Price
MUHAMMAD Jonathan A. C. Brown
MULTICULTURALISM Ali Rattansi
MULTILINGUALISM John C. Maher
MUSIC Nicholas Cook
MUSIC AND TECHNOLOGY Mark Katz
MYTH Robert A. Segal
NANOTECHNOLOGY Philip Moriarty
NAPOLEON David A. Bell
THE NAPOLEONIC WARS Mike Rapport
NATIONALISM Steven Grosby
NATIVE AMERICAN LITERATURE Sean Teuton
NAVIGATION Jim Bennett
NAZI GERMANY Jane Caplan

NEGOTIATION Carrie Menkel-Meadow
NEOLIBERALISM Manfred B. Steger and Ravi K. Roy
NETWORKS Guido Caldarelli and Michele Catanzaro
THE NEW TESTAMENT Luke Timothy Johnson
THE NEW TESTAMENT AS LITERATURE Kyle Keefer
NEWTON Robert Iliffe
NIETZSCHE Michael Tanner
NINETEENTH-CENTURY BRITAIN Christopher Harvie and H. C. G. Matthew
THE NORMAN CONQUEST George Garnett
NORTH AMERICAN INDIANS Theda Perdue and Michael D. Green
NORTHERN IRELAND Marc Mulholland
NOTHING Frank Close
NUCLEAR PHYSICS Frank Close
NUCLEAR POWER Maxwell Irvine
NUCLEAR WEAPONS Joseph M. Siracusa
NUMBER THEORY Robin Wilson
NUMBERS Peter M. Higgins
NUTRITION David A. Bender
OBJECTIVITY Stephen Gaukroger
OBSERVATIONAL ASTRONOMY Geoff Cottrell
OCEANS Dorrik Stow
THE OLD TESTAMENT Michael D. Coogan
THE ORCHESTRA D. Kern Holoman
ORGANIC CHEMISTRY Graham Patrick
ORGANIZATIONS Mary Jo Hatch
ORGANIZED CRIME Georgios A. Antonopoulos and Georgios Papanicolaou
ORTHODOX CHRISTIANITY A. Edward Siecienski
OVID Llewelyn Morgan
PAGANISM Owen Davies
PAKISTAN Pippa Virdee
THE PALESTINIAN-ISRAELI CONFLICT Martin Bunton
PANDEMICS Christian W. McMillen
PARTICLE PHYSICS Frank Close
PAUL E. P. Sanders
IVAN PAVLOV Daniel P. Todes
PEACE Oliver P. Richmond
PENTECOSTALISM William K. Kay
PERCEPTION Brian Rogers
THE PERIODIC TABLE Eric R. Scerri
PHILOSOPHICAL METHOD Timothy Williamson
PHILOSOPHY Edward Craig
PHILOSOPHY IN THE ISLAMIC WORLD Peter Adamson
PHILOSOPHY OF BIOLOGY Samir Okasha
PHILOSOPHY OF LAW Raymond Wacks
PHILOSOPHY OF MIND Barbara Gail Montero
PHILOSOPHY OF PHYSICS David Wallace
PHILOSOPHY OF SCIENCE Samir Okasha
PHILOSOPHY OF RELIGION Tim Bayne
PHOTOGRAPHY Steve Edwards
PHYSICAL CHEMISTRY Peter Atkins
PHYSICS Sidney Perkowitz
PILGRIMAGE Ian Reader
PLAGUE Paul Slack
PLANETARY SYSTEMS Raymond T. Pierrehumbert
PLANETS David A. Rothery
PLANTS Timothy Walker
PLATE TECTONICS Peter Molnar
PLATO Julia Annas
POETRY Bernard O'Donoghue
POLITICAL PHILOSOPHY David Miller
POLITICS Kenneth Minogue
POLYGAMY Sarah M. S. Pearsall
POPULISM Cas Mudde and Cristóbal Rovira Kaltwasser
POSTCOLONIALISM Robert J. C. Young
POSTMODERNISM Christopher Butler
POSTSTRUCTURALISM Catherine Belsey
POVERTY Philip N. Jefferson
PREHISTORY Chris Gosden
PRESOCRATIC PHILOSOPHY Catherine Osborne
PRIVACY Raymond Wacks

PROBABILITY John Haigh
PROGRESSIVISM Walter Nugent
PROHIBITION W. J. Rorabaugh
PROJECTS Andrew Davies
PROTESTANTISM Mark A. Noll
PSEUDOSCIENCE Michael D. Gordin.
PSYCHIATRY Tom Burns
PSYCHOANALYSIS Daniel Pick
PSYCHOLOGY Gillian Butler and Freda McManus
PSYCHOLOGY OF MUSIC Elizabeth Hellmuth Margulis
PSYCHOPATHY Essi Viding
PSYCHOTHERAPY Tom Burns and Eva Burns-Lundgren
PUBLIC ADMINISTRATION Stella Z. Theodoulou and Ravi K. Roy
PUBLIC HEALTH Virginia Berridge
PURITANISM Francis J. Bremer
THE QUAKERS Pink Dandelion
QUANTUM THEORY John Polkinghorne
RACISM Ali Rattansi
RADIOACTIVITY Claudio Tuniz
RASTAFARI Ennis B. Edmonds
READING Belinda Jack
THE REAGAN REVOLUTION Gil Troy
REALITY Jan Westerhoff
RECONSTRUCTION Allen C. Guelzo
THE REFORMATION Peter Marshall
REFUGEES Gil Loescher
RELATIVITY Russell Stannard
RELIGION Thomas A. Tweed
RELIGION IN AMERICA Timothy Beal
THE RENAISSANCE Jerry Brotton
RENAISSANCE ART Geraldine A. Johnson
RENEWABLE ENERGY Nick Jelley
REPTILES T. S. Kemp
REVOLUTIONS Jack A. Goldstone
RHETORIC Richard Toye
RISK Baruch Fischhoff and John Kadvany
RITUAL Barry Stephenson
RIVERS Nick Middleton
ROBOTICS Alan Winfield
ROCKS Jan Zalasiewicz
ROMAN BRITAIN Peter Salway
THE ROMAN EMPIRE Christopher Kelly
THE ROMAN REPUBLIC David M. Gwynn
ROMANTICISM Michael Ferber
ROUSSEAU Robert Wokler
RUSSELL A. C. Grayling
THE RUSSIAN ECONOMY Richard Connolly
RUSSIAN HISTORY Geoffrey Hosking
RUSSIAN LITERATURE Catriona Kelly
THE RUSSIAN REVOLUTION S. A. Smith
SAINTS Simon Yarrow
SAMURAI Michael Wert
SAVANNAS Peter A. Furley
SCEPTICISM Duncan Pritchard
SCHIZOPHRENIA Chris Frith and Eve Johnstone
SCHOPENHAUER Christopher Janaway
SCIENCE AND RELIGION Thomas Dixon and Adam R. Shapiro
SCIENCE FICTION David Seed
THE SCIENTIFIC REVOLUTION Lawrence M. Principe
SCOTLAND Rab Houston
SECULARISM Andrew Copson
SEXUAL SELECTION Marlene Zuk and Leigh W. Simmons
SEXUALITY Véronique Mottier
WILLIAM SHAKESPEARE Stanley Wells
SHAKESPEARE'S COMEDIES Bart van Es
SHAKESPEARE'S SONNETS AND POEMS Jonathan F. S. Post
SHAKESPEARE'S TRAGEDIES Stanley Wells
GEORGE BERNARD SHAW Christopher Wixson
MARY SHELLEY Charlotte Gordon
THE SHORT STORY Andrew Kahn
SIKHISM Eleanor Nesbitt
SILENT FILM Donna Kornhaber
THE SILK ROAD James A. Millward
SLANG Jonathon Green
SLEEP Steven W. Lockley and Russell G. Foster
SMELL Matthew Cobb
ADAM SMITH Christopher J. Berry
SOCIAL AND CULTURAL ANTHROPOLOGY John Monaghan and Peter Just

SOCIAL PSYCHOLOGY Richard J. Crisp
SOCIAL WORK Sally Holland and Jonathan Scourfield
SOCIALISM Michael Newman
SOCIOLINGUISTICS John Edwards
SOCIOLOGY Steve Bruce
SOCRATES C. C. W. Taylor
SOFT MATTER Tom McLeish
SOUND Mike Goldsmith
SOUTHEAST ASIA James R. Rush
THE SOVIET UNION Stephen Lovell
THE SPANISH CIVIL WAR Helen Graham
SPANISH LITERATURE Jo Labanyi
THE SPARTANS Andrew J. Bayliss
SPINOZA Roger Scruton
SPIRITUALITY Philip Sheldrake
SPORT Mike Cronin
STARS Andrew King
STATISTICS David J. Hand
STEM CELLS Jonathan Slack
STOICISM Brad Inwood
STRUCTURAL ENGINEERING David Blockley
STUART BRITAIN John Morrill
SUBURBS Carl Abbott
THE SUN Philip Judge
SUPERCONDUCTIVITY Stephen Blundell
SUPERSTITION Stuart Vyse
SYMMETRY Ian Stewart
SYNAESTHESIA Julia Simner
SYNTHETIC BIOLOGY Jamie A. Davies
SYSTEMS BIOLOGY Eberhard O. Voit
TAXATION Stephen Smith
TEETH Peter S. Ungar
TERRORISM Charles Townshend
THEATRE Marvin Carlson
THEOLOGY David F. Ford
THINKING AND REASONING Jonathan St B. T. Evans
THOUGHT Tim Bayne
TIBETAN BUDDHISM Matthew T. Kapstein
TIDES David George Bowers and Emyr Martyn Roberts
TIME Jenann Ismael
TOCQUEVILLE Harvey C. Mansfield
LEO TOLSTOY Liza Knapp
TOPOLOGY Richard Earl
TRAGEDY Adrian Poole
TRANSLATION Matthew Reynolds
THE TREATY OF VERSAILLES Michael S. Neiberg
TRIGONOMETRY Glen Van Brummelen
THE TROJAN WAR Eric H. Cline
TRUST Katherine Hawley
THE TUDORS John Guy
TWENTIETH-CENTURY BRITAIN Kenneth O. Morgan
TYPOGRAPHY Paul Luna
THE UNITED NATIONS Jussi M. Hanhimäki
UNIVERSITIES AND COLLEGES David Palfreyman and Paul Temple
THE U.S. CIVIL WAR Louis P. Masur
THE U.S. CONGRESS Donald A. Ritchie
THE U.S. CONSTITUTION David J. Bodenhamer
THE U.S. SUPREME COURT Linda Greenhouse
UTILITARIANISM Katarzyna de Lazari-Radek and Peter Singer
UTOPIANISM Lyman Tower Sargent
VATICAN II Shaun Blanchard and Stephen Bullivant
VETERINARY SCIENCE James Yeates
THE VIKINGS Julian D. Richards
VIOLENCE Philip Dwyer
THE VIRGIN MARY Mary Joan Winn Leith
THE VIRTUES Craig A. Boyd and Kevin Timpe
VIRUSES Dorothy H. Crawford
VOLCANOES Michael J. Branney and Jan Zalasiewicz
VOLTAIRE Nicholas Cronk
WAR AND RELIGION Jolyon Mitchell and Joshua Rey
WAR AND TECHNOLOGY Alex Roland
WATER John Finney
WAVES Mike Goldsmith
WEATHER Storm Dunlop
THE WELFARE STATE David Garland
WITCHCRAFT Malcolm Gaskill
WITTGENSTEIN A. C. Grayling
WORK Stephen Fineman

WORLD MUSIC **Philip Bohlman**
WORLD MYTHOLOGY **David Leeming**
THE WORLD TRADE ORGANIZATION **Amrita Narlikar**
WORLD WAR II **Gerhard L. Weinberg**
WRITING AND SCRIPT **Andrew Robinson**
ZIONISM **Michael Stanislawski**
ÉMILE ZOLA **Brian Nelson**

Available soon:

IMAGINATION **Jennifer Gosetti-Ferencei**
THE VICTORIANS **Martin Hewitt**

For more information visit our website

www.oup.com/vsi/

David W. Macdonald

BIODIVERSITY CONSERVATION

A Very Short Introduction

Great Clarendon Street, Oxford, OX2 6DP,
United Kingdom

Oxford University Press is a department of the University of Oxford.
It furthers the University's objective of excellence in research, scholarship,
and education by publishing worldwide. Oxford is a registered trade mark of
Oxford University Press in the UK and in certain other countries

Published in the United States of America by Oxford University Press
198 Madison Avenue, New York, NY 10016, United States of America

British Library Cataloguing in Publication Data
Data available

Library of Congress Control Number: 2023930670

ISBN 978–0–19–959227–2

Printed and bound by
CPI Group (UK) Ltd, Croydon, CR0 4YY

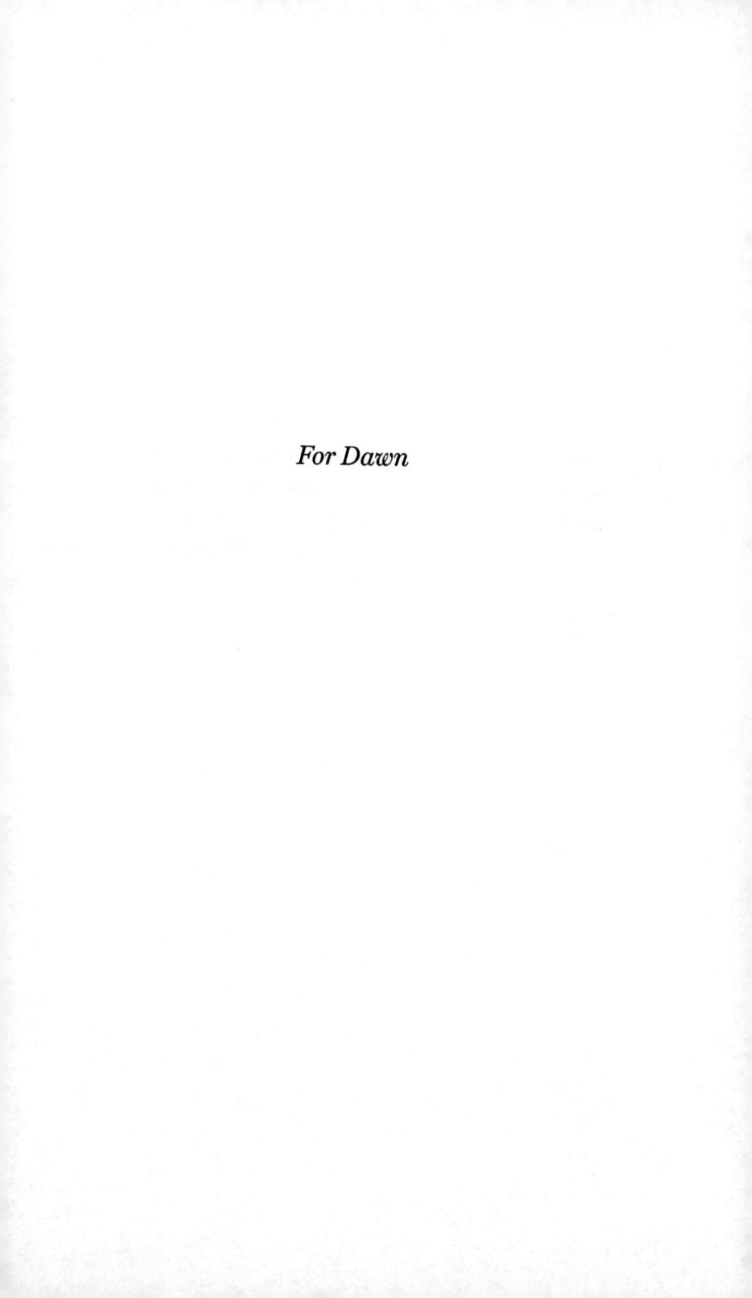

For Dawn

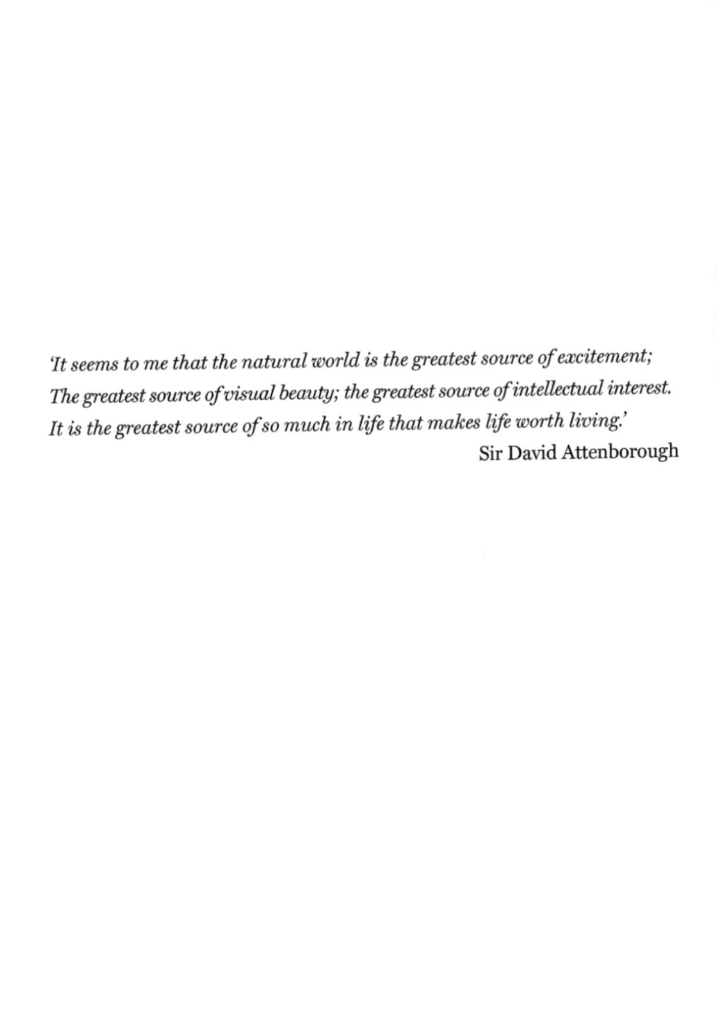

'It seems to me that the natural world is the greatest source of excitement; The greatest source of visual beauty; the greatest source of intellectual interest. It is the greatest source of so much in life that makes life worth living.'

Sir David Attenborough

Contents

Acknowledgements

Almost everything worth saying seems to have been said already by Mark Twain: 'I didn't have time to write a short letter, so I wrote a long one instead.' Having been tasked to write a very short book about a very big topic, I can confirm it takes time. Further, as readers will learn within, biodiversity is so diverse that an egalitarian book, especially a very short one, could devote less than a punctuation mark to each species, so inevitably I have fallen prey to bias: my examples draw disproportionately on vertebrates, especially mammals, for which I don't much apologize—they illustrate well the problems and principles of biodiversity conservation, which is what matters. I also draw more than randomly on the work of the Wildlife Conservation Research Unit (WildCRU) at the University of Oxford for the good reasons that, as its founding Director, I know it best, and its mission 'to achieve practical solutions to conservation problems through original scientific research' goes to the heart of the topic of this book. Even the examples I have mentioned generate a bibliography too vast to include: for those hungry for the full story the references are given at www.wildcru.org/VSI_Biodiversity_Conservation/.

Above all I am grateful to my wife Dr Dawn Burnham—who has read, shaken, and stirred every word—for her professional insights and personal support. I also thank particularly my friend and colleague Dr Christopher O'Kane for his unfailing effectiveness (made possible by Peter and Gyongyver Kadas). In addition to Dawn and Christopher, the whole book was fruitfully criticized by risen conservation stars Drs Darragh Hare and Laura Perry, and by rising representatives of the next generation, Chrishen Gomez and Claire Marr, to whom I'm limitlessly grateful, along with John Salmon who also read the whole manuscript as my representative normal reader (if that's an apt description of a top surgeon). To these stalwarts, I add thanks to specialist critics of individual chapters, Dr Luca Chiaverini, Dr Alayne Cotterill, Professor Chris Dye, Dr Kim Jacobsen, Dr Kerry Kilshaw, Professor Andrew Loveridge, Dr Ewan Macdonald, Dr Silvio Marchini, Dr Axel Moehrenschlager, Dr Chris Newman, Dr Ugyen Penjor, Professor Alex Teytelboym, Dr Peter Tyrrell, and others of the extraordinary WildCRU diaspora (all of whom, but none more than me, have been enabled by the remarkable support of Tom and Dafna Kaplan).

I am so pleased to be years late in delivering this book. Had I been punctual I'd have pre-empted biodiversity conservation's formative growth spurt that has left the subject almost unrecognizably metamorphosed from its persona just a few years ago (although I haven't forgotten the help of Drs Ros Shaw and Ruth Feber with an earlier instar). My relief at not writing the book while the topic was still larval is tempered only by my apology for the frustration I caused to OUP's Latha Menon, the series' mother. I am particularly grateful to her for providing me with comfort that the era of editors that edit is not entirely past.

Readers will see, in these pages, just how radical have been the accelerating impacts on biodiversity and biodiversity conservation, within scarcely a professional lifetime. I began my career in Borneo, and so too did my son, Ewan, but between our generations a third of the forest had gone. A vocation once rewarded by reading a pawprint in the mud, or by staring deeply into the eyes of another species, or even getting sufficiently into its skin to know what it would do next, and to understand why, arrives now at the shuddering intersection of teragrams of carbon, global pollutants, viral genomes, market forces, property rights, and geopolitics, with its meta-analyses and models. There is both thrill and chill in this new reality. Twenty years ago the present would have been barely imaginable, so writing now, on her first birthday, I can only wonder how things will seem when my granddaughter, Hannah, is old enough to read this book. Wherever biodiversity conservation's journey from groundedness to geopolitics may go next, it is perilously urgent.

List of illustrations

Part 1

Setting the scene

Chapter 1
What is biodiversity, and why does it matter?

Biodiversity is a term that embraces the diversity of life at different scales, and is almost, but importantly not quite, colloquially synonymous with nature or wildlife. Biodiversity turns out to be crucial to safeguarding human well-being, as well as the well-being of nature and all its moving parts. This chapter answers the questions 'what is biodiversity?' and 'why does it matter?' by exploring how biodiversity is assembled as the living cogs of nature's machinery. Chapter 2, 'What's the problem?', tackles the questions of what dictates whether biodiversity is resilient and, where it is frail, what threatens it (and thus all of us). Next, what's to be done about it? Or, as expressed in Chapter 3: what is the purpose of biodiversity conservation?—a question that is nowhere near as easy to answer as it might once have seemed. From those foundations, the rest of this *Very Short Introduction* will dissect some of these risks further, consider what can be done about them, why that matters, and what's next.

The astonishing diversity of life on Earth

You are a eukaryote; that is an organism made up of cells that possess a membrane-bound nucleus that holds genetic material as well as membrane-bound organelles. Prokaryotes are organisms whose cells have neither nucleus nor membrane-encased organelles. In the more than 250 years since Swedish biologist

Carl Linnaeus began the science of taxonomy, 1.2 million eukaryotic (animals, plants, fungi, and protists) species have been identified and classified—less than 15 per cent of 8.7 million eukaryotic species estimated to exist. On an average day 50 new species, most commonly insects, are formally described—even amongst mammals a new species is discovered roughly every three years: the most recent are Rice's whale (2021) and the Popa langur (2020) (Figure 1). A new large vertebrate from the open ocean is found every five years.

If the world's biodiversity is a library, how is it indexed? A comprehensive answer is given in Peter Holland's *Very Short Introduction* (VSI) to the Animal Kingdom, with parallel diversity

1. The Popa langur. A new species of monkey living on an extinct volcano in Myanmar has been described for the first time in 2020. There are thought to be only around 200–250 individual Popa langurs left, meaning they are already considered critically endangered.

in the Plant Kingdom (Tim Walker's *Plants* VSI). Amongst contending definitions of biodiversity, the *Oxford Dictionary*'s is a straightforward start: 'the existence of a large number of different kinds of animals and plants which make a balanced environment'. The American Museum of Natural History elaborates that biodiversity 'refers to the variety of life on Earth at all its levels, from genes to ecosystems, and can encompass the evolutionary, ecological, and cultural processes that sustain life'. Nowadays, measuring genetic diversity is within our grasp—I just received from the Sanger Institute the complete genome of one of the badgers I've been studying. Not long ago that would have seemed like science fiction, yet already it's possible to buy a species' genome for the price of a restaurant meal. As for diversity and abundance at the level of populations, that too can be measured.

Taxonomically, individuals are indexed by category, each tracing back to progressively more distant common evolutionary ancestors. The categories are named: kingdom, phylum, class, order, family, genus, species (there are also many sub-categories). For example, a chimpanzee would be filed under: Kingdom: *Animalia*; Phylum: *Chordata*; Class: *Mammalia*; Order: *Primates*; Family: *Hominidae*; Genus: *Pan*; Species: *P. troglodytes*. You, by the way, would only differ from the chimpanzee at the genus level (*Homo*) and if you were to look in the library's archives you'd find nine human species walked the Earth 300,000 years ago, all of whom were humans (*Homo* but not *H. sapiens*), and many of whom were most likely driven extinct by our ancestors through competition (with a bit of genetic blurring of boundaries by interbreeding thrown in—you are probably 2 per cent Neanderthal). Just for animals there are 33 phyla, 107 classes, ~440 orders, ~6,000 families, and ~110,000 genera encompassing microscopic invertebrates, such as the 0.1-mm-long tardigrade or 'water bear', to whales. The two largest Orders of the Class Mammalia are the bats (over 1,400 species) and rodents (about 1,500 species). For plants the spectrum runs from the watermeal, a flowering plant 0.1 mm in diameter, to the giant sequoia tree, found in California with trunk diameters of almost 8 metres. There are 14 phyla in the plant kingdom, 8 phyla of fungi, and 41 accepted bacterial phyla (a tally expected eventually to total 1,300). The span of genetic diversity found within one bacterial species *Legionella pneumophila* is as great as the genetic distance between mammals and fish.

The span of diversity that often springs to mind, for example amongst land mammals from the 2 g Kitti's hog-nosed bat to the 6,000 kg African elephant, is spectacular as a homage to the power of adaptation through natural selection; however, it is but a drop in the ocean of the full panoply of biodiversity. Parasites comprise a hefty proportion of global biodiversity—each species of animal or plant generally coexists with at least one each of specialist macroparasite and microparasite, and may support a whole community of parasites. Continuing discoveries of 'extremophile' organisms deep in the soil, in underground lakes, and in oceanic vents suggest that unknown unknowns remain to be discovered.

Beyond numbers, it is the range of life on Earth that truly astounds. Whether in size, from tiny Phytoplankton (weighing only 1×10^{-11} g) that form the base of oceanic food webs and produce much of the Earth's oxygen, to the 200-tonne blue whale, the largest animal ever to have existed on Earth. Or in adaptation to a particular niche, from the fig wasp's reproductive cycle inside flowering figs, to the giant panda that eats only bamboo. Or in evolutionary pathway, from the eight-limbed and intelligent octopus to ourselves. The diversity and intricacy of the adaptation of Earth's inhabitants are marvels whose richness we scarcely begin to understand.

How is biodiversity assembled?

In 1957 ecologist Evelyn Hutchinson crystallized the notion of the niche (as an '*n*-dimensional hyperspace') to characterize and quantify the positioning of each species in its environment.

Nowadays, biologists appreciate that multiple species assemble into communities (Table 1), each using the dimensions of their environment in subtly different ways that enable them to coexist.

Patterns of co-occurrence might reflect shared habitat requirements, be driven by biotic interactions, or be a consequence of human-made—anthropogenic—pressures, such as changes in land-use. Two co-occurring species might be predator and prey, whose fates interweave according to the relationship of life and dinner, or they might be rivals. Where two seemingly similar species live side by side in the same habitat—'coexist sympatrically'—they may avoid each other in their use of space and time. Species' niches bump up against each other in highly dynamic ways. For example, from a spotted hyaena's point of view, encounters with lions swing from being positive during the wet season, when they meet lions while scavenging their prey, to negative during the dry season, when their encounters generally involve lions driving them away from remaining water holes, where scarce prey concentrate.

Amongst mammals, guilds of carnivores illustrate the drivers of community structure: for example, character displacement, where the evolution of different forms indicates different functions that allow coexistence. Niche partitioning occurs when species are different enough to coexist—weasels, stoats, polecats, mink, martens, fishers, and sable. Competing species face the cut-and-thrust of intra-guild hostility, which often takes the form of a bigger species, say tigers, bullying a smaller one, leopards. Many carnivores are apex predators—that is, they perch aloft the food pyramid—and as such play important roles in ecosystem functioning, providing ecological stability by operating as a negative feedback on prey populations.

Members of the cat family, the Felidae, offer a particularly useful model for understanding niche separation, competition, and drivers of biodiversity. A recent study in the Northern Forest

Table 1 Levels of organization in nature

Genes	Individuals	Populations	Species	Guilds	Assemblages	Communities	Ecosystem	Biosphere
The basic physical and functional unit of heredity	One organism and also one type of organism	A subset of individuals of one species that occupies a particular geographic area and, in sexually reproducing species, interbreeds	A group of organisms that can reproduce naturally with one another and create fertile offspring	A group of species that exploit the same class of environmental resources in a similar way	Taxonomically related group of species populations that occur together in space	All of the populations of different species that live in the same area and interact with one another	An ecological community comprising biological, physical, and chemical components, considered as a unit	The parts of Earth where life exists—all ecosystems

Complex of Myanmar used camera-traps to explore how tigers, clouded leopard, Asiatic golden cat, marbled cat, and leopard cat all occur together. Tiger and marbled cats were primarily diurnal, clouded leopard and leopard cat were nocturnal, and golden cat exhibited round-the-clock (cathemeral) activity. Amongst the three medium-sized species (clouded leopards, golden cats, and marbled cats), the greater the similarity in their body size, the less they used the same space at the same time. It was amongst the three smallest species that the largest differences in use of space and time occurred. These small cats slot their lives around those of their large tyrannical cousins; the marbled cats (3 kg) especially avoided the Asiatic golden cat (8 kg), which, in turn, were especially avoided by the (4 kg) leopard cats. These insights into a carnivore guild assembly reflect, in microcosm, principles of competition and coexistence that reverberate, albeit through complicated larger-scale processes of habitat heterogeneity, throughout the animal and plant kingdoms.

One mechanism by which species evolve to coexist is adaptive radiation—the production from one species, often rather quickly, of many species occupying different ecological niches. The classic, enduringly vivid, example of adaptive radiation is provided by the 14 species of finches that carved up the Galápagos archipelago between them over the last 2–3 million years. As Darwin wrote in 1842,

> The most curious fact is the perfect gradation in the size of the beaks of the different species of Geospiza.... Seeing this gradation and diversity of structure in one small, intimately related group of birds, one might fancy that, from an original paucity of birds in this archipelago, one species has been taken and modified for different ends.

He was spot on. Amongst this guild, competition and coexistence are reflected in the association between beak size, beak structure, and diet—most obvious when comparing the insectivorous small warbler finches (about 8 g) and the granivorous large ground finch (30 g).

At a smaller scale, the four species of ants linked to one species of acacia tree, *Acacia drepanolobium*, form an invertebrate guild. Near the nutrient-enriched soil that termites create in and around their mounds, more fresh Acacia shoots sprout and densities of litter-dwelling invertebrates (an important food source of the acacia ants) are higher than further away from the mounds. This spatial variation in resource availability is correlated with competition amongst the ants for host trees; near the mounds, competitively dominant species are more likely to supplant subordinate species, whilst subordinates replacing dominants from host trees increases with distance from termite mounds. The termite-induced heterogeneity in habitat thus influences the dynamics of the acacia ant community, contributing to species coexistence in an intensely competitive community.

How large numbers of competing plant species manage to coexist continues to be a conundrum—the classical explanation, that each species occupies its own niche, seems to creak under the weight of observation that most plants require the same set of resources and have a limited number of ways to acquire them. Nonetheless, delving into fine detail reveals guilds of plants segregated along such environmental niche axes as gradients of light, soil moisture, and root depth, and also partitioning soil nutrients with the helpful mediation of microbes in the soil. Indeed, recent research reveals how plants are capable of both actively promoting microbial populations and coordinating their engagement with microbes to optimize nitrogen and phosphorus uptake.

Graduating from species and guilds, the next level of organization is assemblages—for example the 51 different species of insectivorous bats captured in approximately 3 km^2 of primary dipterocarp rainforest in Malaysia, or the entire range of more than 300 fish species found in coastal waters around the British Isles.

The power of adaptation is astonishing and the relationships that link the elements of biodiversity are close to infinite in their intricacy. Consider the bizarre partnership that allows the dotted humming frog to nestle safe from its predators between the hairy limbs of the burrowing tarantula (the frog eats ants that threaten the spider's eggs); or the hornbills and dwarf mongooses who trade benefits of foraging and vigilance; or how the digging out of clams by sea otters leads to higher genetic diversity in the fields of underwater eelgrass disturbed by this excavation. The intricate and far-reaching relationships of biodiversity, and the complexity they pose to conservation, are emphasized in John Vucetich's book *Restoring the Balance*—heavy infestations of diminutive ticks negatively impact moose survival, reverberating through to forest structure and wolf population dynamics. It is these relationships that are the linkages in the ecosystem processes, which lead to ecosystem services, large and small, on which the human enterprise depends, and which are played out on the stages of different biomes.

Biomes

Biomes are the regions of Earth that can be distinguished by their climate, fauna, and flora, with the organisms that live in each biome adapted to its circumstances, in particular to the climate and vegetation type. There are five major types of biome (Figure 2)—aquatic, grassland, forest, desert, and tundra—each of which can be further divided according to scholarly taste.

Biodiversity differs in different biomes, and can be measured at different scales (within a patch, between patches, and across a landscape—called, respectively, alpha, beta, and gamma diversity). Forests cover about one-third of the land, are dominated by trees, and contain much of the world's terrestrial biodiversity, including insects, birds, and mammals. The three major forest biomes are boreal, temperate, and tropical, each occurring at different latitudes, and therefore experiencing different climates. Tropical

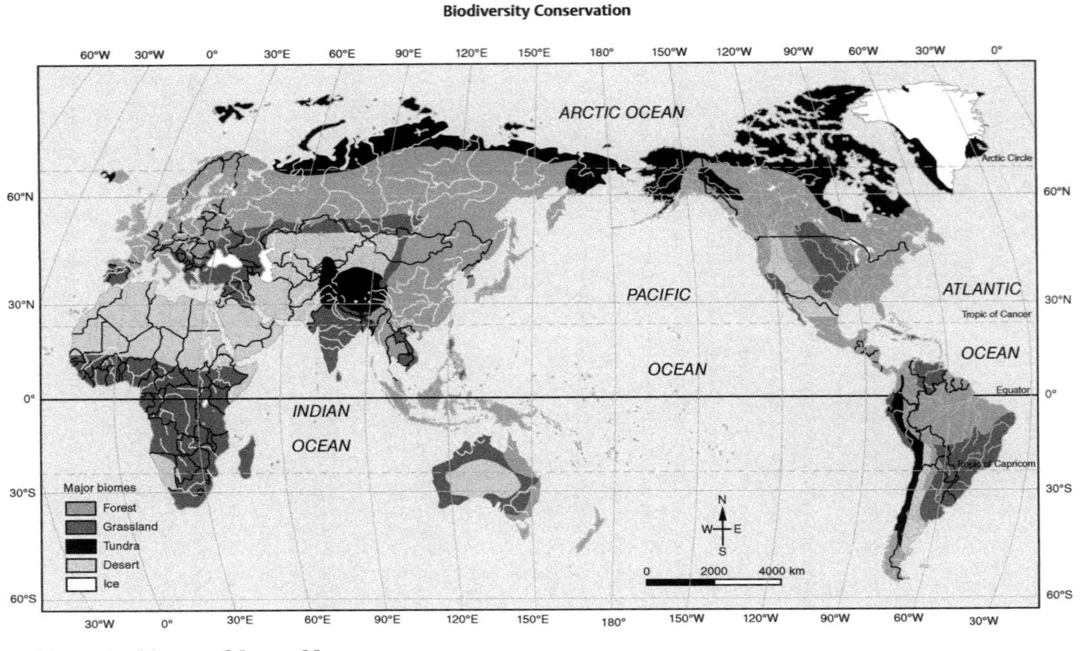

2. The major biomes of the world.

forests are warm, humid, and found close to the equator and they contain the greatest biodiversity; temperate forests are found at higher latitudes and experience all four seasons, whilst boreal forests are found at even higher latitudes, have the coldest and driest climate, and consequently the lowest biodiversity.

Grasslands, a transition on the spectrum from forest to desert, are open regions dominated by grass that have a warm, dry climate, and of course are mainly inhabited by grazing animals—pastoralist peoples value them greatly for livestock grazing. Neither specialized grazers (voles or horses), nor their specialist predators (weasels or lions) existed until about 55 million years ago because it was only then that evolution and earlier climate change conspired for grass to become abundant. There are two types of grasslands: savannahs and temperate grasslands. Savannahs occur closer to the equator, are peppered with trees, and cover almost half of Africa—our ancestors evolved on them. Temperate grasslands (e.g. steppes and prairies) occur further from the equator, lack trees or shrubs, receive less precipitation, and, consequently, display lower levels of biodiversity than savannahs.

Deserts cover around 20 per cent of Earth's surface, are always dry (< 50 cm rainfall p.a.), can be either cold or hot, but are mostly subtropical. The natural tally of biodiversity in deserts can be low, consisting particularly of small mammals, reptiles, and invertebrates. Least hospitable is the tundra biome: from –34 to 12°C, and only 15–25 cm precipitation per year, poor soil nutrients, and short summers. Tundra vegetation is simple, including shrubs, grasses, mosses, and lichens, partly due to the permafrost (a frozen layer under the soil surface), and biodiversity is very low.

Aquatic biomes include freshwater and marine, the latter covering almost three-quarters of Earth's surface and containing the greatest biodiversity of any biome. Whilst terrestrial organisms are generally limited by temperature and moisture, their aquatic counterparts are principally limited by sunlight and dissolved nutrients. Despite its fluidity, seawater is stratified and so is its biodiversity: there is sufficient sunlight for photosynthesis in only the top 200 m, and dead organisms sink, to be decomposed at great depths, so deep water contains more nutrients than surface water.

James Lovelock argued that the tight coupling between biodiversity and the Earth's abiotic systems creates a whole—Gaia—that is self-regulating. His idea resonates in the 2001 Declaration of Amsterdam, which concluded 'The Earth System behaves as a single, self-regulating system with physical, chemical, biological and human components.'

Biodiversity hotspots

The appreciation that there is more biodiversity in some places than others triggered the thought by another pioneering conservationist, Norman Myers, that we might identify the richest concentrations—he called them biodiversity hotspots—which would disproportionately repay the effort of conservation. Originally 25 biodiversity hotspots were identified globally (Figure 3). They collectively support 44 per cent and 35 per cent of the world's vascular plants and terrestrial vertebrates, respectively, in a tiny area equal to 1.4 per cent of the Earth's land surface.

Not only are some places, the hotspots, richer in, and more representative of, the biodiversity characteristic of their region and biome than others, but amongst these hotspots some are particularly at risk. For example, rates of biodiversity loss in South-East Asia are among the highest in the world, and the Indo-Burma and South-Central China biodiversity hotspots rank among the world's most threatened. The camera-trapping grids that revealed the cat guild in Myanmar were part of a much larger coordinated study with cameras at over 1,000 locations in 15

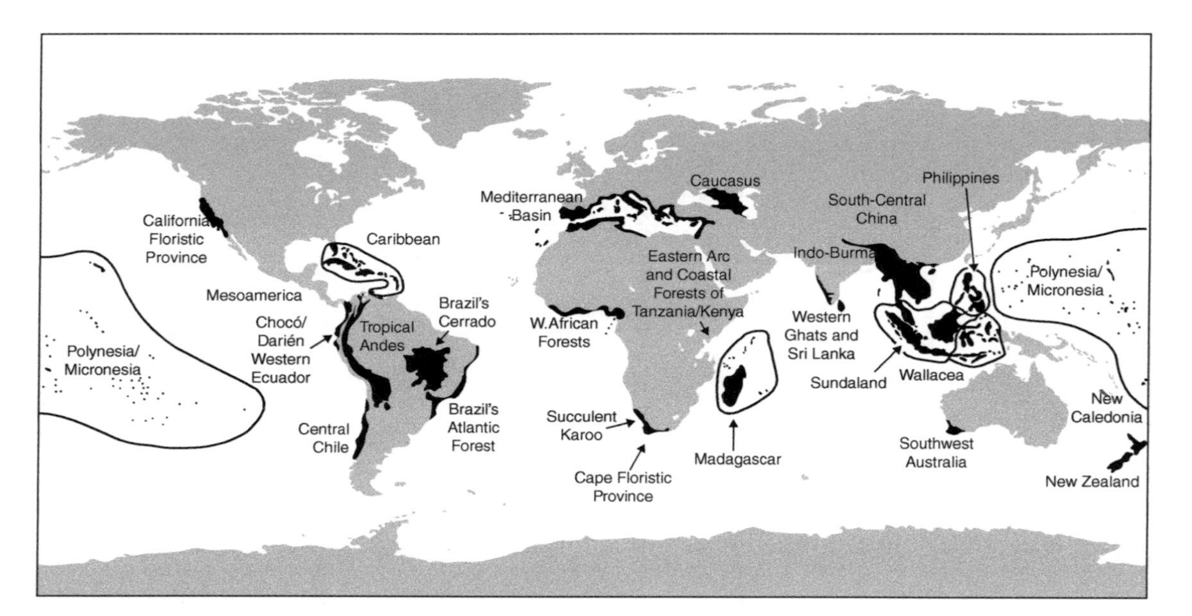

3. **Biodiversity hotspots situated within high (black) biodiversity areas.**

landscapes across seven mainland countries in South-East Asia from Nepal to Malaysia. These cameras recorded 90 different vertebrate species. This was a big sample of regional forest biodiversity, providing the opportunity to ask where this aspect of biodiversity was most diverse, what determined its richness, and how the answer mapped onto where protected areas like national parks have been established. These questions require investigating lots of different variables, each at several different scales (ranging between 250 m and 32 km). This revealed two main biodiverse centres: the Thai-Malay Peninsula and the mountainous region of south-west China (Figure 4a).

It was a vast undertaking to record that fraction of biodiversity that exposed itself to the camera-traps, so it makes sense to ask, for next time, if there's a short-cut indicator. It turns out there is: small and medium-sized carnivores were the strongest indicators of species richness, whilst the best predictors of the greatest biodiversity were latitude and elevation. Sadly, this mapping of observed and extrapolated forest biodiversity across South-East Asia poorly matched the current system of protected areas. The majority of areas supporting the highest predicted biodiversity are not formally protected (Figure 4b). These findings highlight areas of high priority for biodiversity conservation, and provide a framework for prioritizing conservation areas for multiple species across international boundaries. Although it's important to be concerned about the threat to particular species, multi-species conservation approaches are more efficient, better value for money, and are more likely to deliver and protect ecosystem processes.

Large-scale ecosystem services underpinned by biodiversity

Ecosystem Services are benefits people obtain from ecosystems, and they have been broadly categorized in the Millennium Ecosystem Assessment of 2005:

(a)

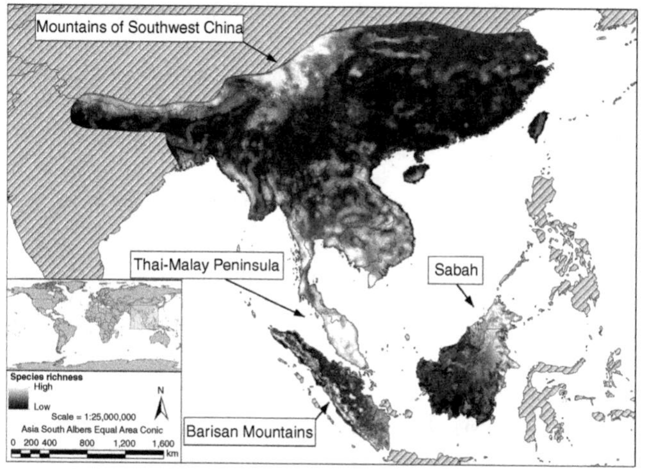

(b)

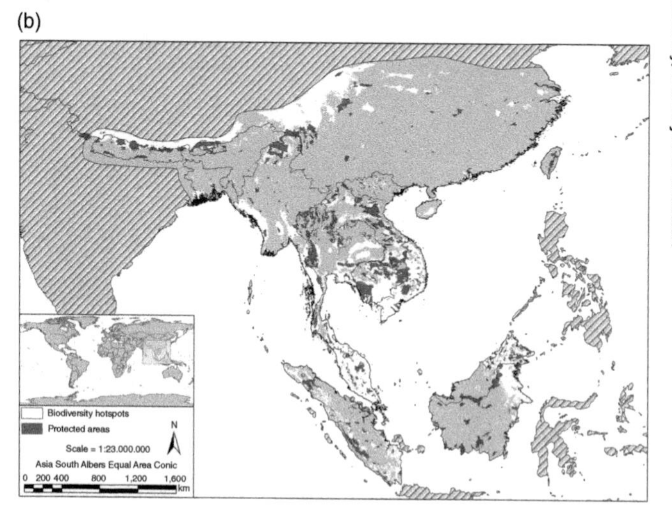

4. (a) Predictive map of species richness in mainland South-East Asia, Borneo, and Sumatra: paler areas are the more biodiverse (shading depicts relative species richness; maximum species richness is lower on the two islands than the mainland); (b) the locations of the same (white shaded) hotspots depicted relative to the locations of current protected areas.

(i) Provisioning Services such as supplies of food, fresh water, fuel, fibre, and other goods;
(ii) Regulating Services such as climate, water, and disease regulation as well as pollination;
(iii) Supporting Services such as soil formation and nutrient cycling; and
(iv) Cultural Services such as educational, aesthetic, and cultural heritage values as well as recreation and tourism.

One way or another, biodiversity contributes to all these services, including the carbon and nitrogen cycle, at an estimated value of US$125–40 trillion per year—a sum, according to the OECD in 2019, equivalent to more than one and a half times global GDP. For this, generally without paying for it, humanity gets from biodiversity the formation of soils, the provision of food and fibre, air quality and climate regulation, the regulation of water supply and quality, along with physical and mental health and the cultural and aesthetic value of plants and animals.

Chapter 2
What's the problem?

Species face two sorts of problems: numbers can decline and distributions can shrink. Either or both can cause the harmony of biodiversity's processes to become discordant. That jolt to community composition, and its cascading consequences for ecosystem functioning, can drive species towards rarity and culminate in extinction. It is amongst the tasks of biodiversity conservation to document these changes and their consequences, to understand and hence anticipate them, and to intervene to prevent, mitigate, and sometimes to remediate them.

The human impact on life on Earth has increased considerably not only since the agricultural and industrial revolutions, but exponentially during my lifetime. Indeed, even since 1970, burgeoning industrialization, pollution, and resource use have multiplied human impacts on the biosphere during what is ominously termed the 'great acceleration'. The damage to biodiversity is ultimately driven by the demands of a growing human population with a rising per capita income and consequently heavier consumption. Nature is supplying more materials than ever before, but this has come at the cost of unprecedented global declines in the extent and integrity of ecosystems, distinctness of local ecological communities, and the abundance of wild species. This is mainly due to habitat loss and degradation as natural habitats are converted for agriculture and forestry. Despite incontestable evidence linking humanity's well-being to that of nature—we are all in this together—a generation of dedicated efforts to conserve biodiversity has scarcely staunched its loss or slowed the degradation of the natural processes that it powers. The demand for land to produce food, animal feed, and energy has increased rapidly, putting at risk numerous ecosystem services upon which humanity relies: one species knowingly sealing the fates of the myriad others. Prior to the Anthropocene—the epoch during which humans have unsustainably impacted nature—less than 1 of every 1,000 mammal species went extinct every millennium. Nowadays, extrapolations from known species groups estimate current extinction rates to be about 100 to 1,000 times faster. This brings us precariously close to a sixth mass extinction. (It is small solace that the losses to several taxa don't quite meet the criterion of a mass extinction at 75 per cent: 41 per cent for amphibians, 63 per cent for cycads.)

Consider the African lion. We might not be surprised if this top predator is at particular risk, although it would be shocking if a species so iconic and so globally revered could be allowed to become imperilled. But that is exactly what has happened, at disquieting speed: a geographic range retraction of ~85 per cent since AD 1500, and a decline of ~75 per cent during the last five decades (Figure 5). That is, in recent decades we have lost an average of ~1,400 lions each year, and now the range of the survivors is so fragmented that only 20 of 34 sub-Saharan countries retain core lion habitat.

Worse yet, over the previous three lion generations—21 years—47 monitored lion populations had declined by an average of 43 per cent, suggesting that as few as about 20,000 individuals survive. That decline has been worst in West/Central Africa at 66 per cent, and not much better in East Africa, at 59 per cent.

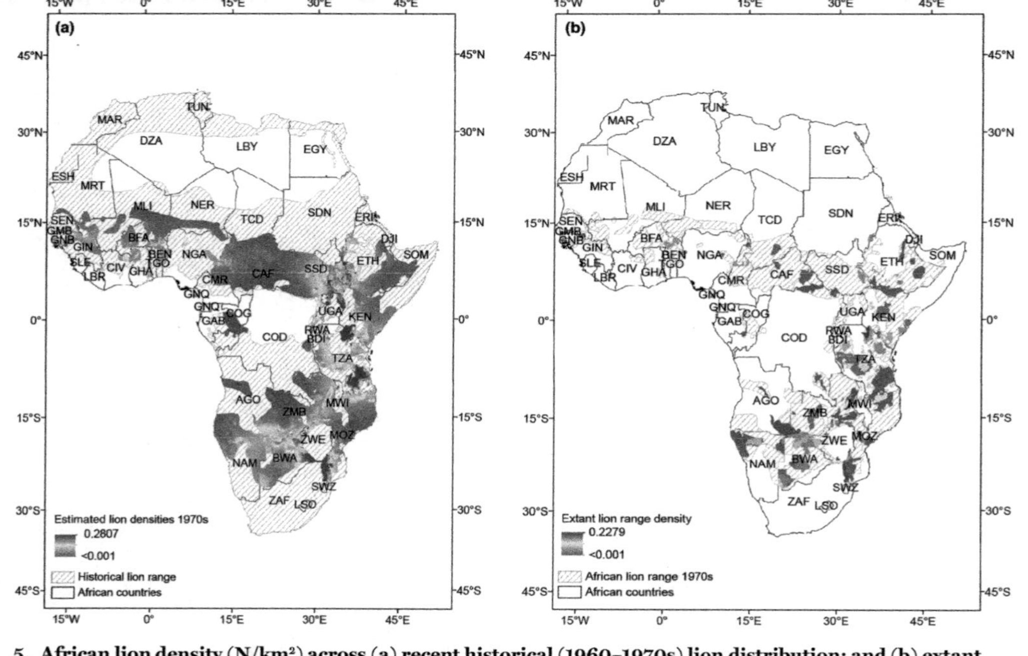

5. African lion density (N/km²) across (a) recent historical (1960–1970s) lion distribution; and (b) extant range showing lion population densities.

6. Cecil the lion.

Such is the plight of lions which, as the King of Beasts, might claim particular attention, but a more egalitarian view of biodiversity would recognize that lions are one of about 1.8 million known species of animal (Figure 6). Thus in this book, amongst the democracy of species lions might expect to be allocated about 0.02 of a word. Attending fairly to the estimated 9 million species of plants (391,000 are described) would, all else being equal, further reduce the lion's fair share of attention to a minute fraction of a full stop (that's before considering the estimated 1 trillion species of microbes).

Although the shrinkage and degradation of biodiversity may cumulatively have the greater effect, the most strident alarm call for biodiversity conservation comes in the all-or-nothing metric of extinction. The International Union for Conservation of Nature (IUCN) Red List of Threatened Species uses experts to assess the threats to species according to a set of criteria (<www.iucnredlist.org/>). Of 76,000 assessed by the IUCN, 40 per cent are threatened (furthermore 15 per cent of mammal species, 24 per cent of amphibian species, and 29 per cent of marine species remain worryingly 'Data Deficient').

A collaboration between 210 scientists from 42 countries, led in 2020 by the Royal Botanic Garden, Kew, concluded that 40 per cent of plant species are at risk of extinction. Cycads—'living fossil' seed plants such as the sago palm, that existed before dinosaurs—are the most threatened (63 per cent). The causes are a familiar dirge: the destruction of wild habitat to create farmland, over-harvesting, building, invasive species, pollution, and climate change; and plant food, the detritus of tiny lives accumulated over millennia in the form of soil organic matter, is being washed away. Of plants used as herbal medicines, 723 species are threatened with extinction. A 2022 analysis used Artificial Intelligence to reveal that more than 1,000 species of palms are at risk of extinction.

Their dwindling numbers squashed on car windscreens provide a ready measure of the decline in insects. Over almost three decades there has been a 76 per cent decline in flying insect biomass across 63 protected areas in Germany. The general decline in flying insects is probably affecting pollination, soil fertility, nutrient recycling, predation, and herbivory. Only 20 per cent of insect species have been described, so it's urgent to discover how they, and the undescribed majority, contribute to ecosystem services.

By the 1960s, powerful trawlers equipped with radar, electronic navigation systems, and sonar revolutionized fishing, leading to an object lesson in unsustainable use. By 1992, Northern Cod populations had fallen to 1 per cent of historical levels. The stability of marine ecosystems was further undermined by huge catches of non-commercial fish. Lessons were not learned: slight upturns in fish stocks were speedily reversed by gold-rush greed, while increasing oceanic temperatures reduced productivity and hence fish stocks. Regulation should focus on 'Fisheries-conservation hotspots', diagnosed by the combination of increasing exploitation, high biodiversity, and poor conservation. Encouragingly, in the north-east Arctic, where cod fishing was curtailed and the size of the fleet reduced, cod are starting to thrive again and they have even begun to mature at a later age.

Since the 1980s amphibian populations have plummeted: in 2021 over 40 per cent of amphibian species were IUCN Red Listed. A major cause is chytrid fungi, which emerged in the Korean peninsula in the early 20th century, transported worldwide by expanding global trade in amphibians. Diseases, habitat destruction and modification, exploitation, pollution, pesticide use, introduced species, and perhaps ultraviolet-B radiation are resulting in amphibian extinctions at 25,000–45,000 times the background rate.

Of ~40 per cent of reptiles Red List assessed (compared with ~99 per cent of birds and mammals), 20 per cent are threatened with extinction. The main problem is habitat modification, due to mining, agriculture, grazing, plantations, and patch size reduction, with the sinister background drumbeat of invasive species, pollution, disease, unsustainable use, global climate change, and the wildlife trade. An estimated 35 per cent of reptile species are traded online, 75 per cent of traded species are not covered by international trade regulation, and 90 per cent of traded reptile species and half of traded individuals are taken from the wild.

Birds are in similarly steep decline. In North America bird abundance is down by 29 per cent (3 billion fewer birds) since 1970. The network of weather radar stations reveals a steep decline in biomass passage of migrating birds. Habitat destruction and degradation affects 93 per cent of threatened birds (Figure 7). Another group of flying vertebrates, bats, is similarly in decline—sometimes dramatically: white-nose syndrome has killed over 90 per cent of northern long-eared, little brown, and tri-coloured bat populations in fewer than 10 years. More happily, of the 11 bat species monitored in the UK, population trends of five are increasing.

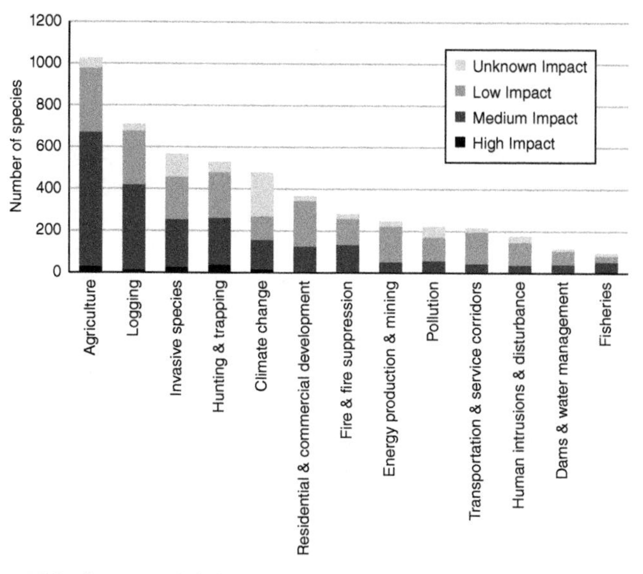

7. Main threats to globally threatened birds.

In 2021 the IUCN considered 26 per cent of mammal species to be at risk of extinction. Dramatic examples arise directly or indirectly from pounding under the human footprint. Directly, in both Laos and Cambodia, over a decade of camera-trapping, my own teams were dismayed to document the precipitous decline to zero of Indochinese leopards, due to illegal snaring. Indirectly, having flourished throughout Britain during the 10,000 or so years prior to the 1930s, water voles were reduced to 2 per cent of their former number by the combined effects of habitat loss and invasive mink. Beware the shifting baseline: modern dismay at declining whitetip and silky sharks is blinkered to the fact that they had already declined by 99 per cent since the 1950s.

This audit of losses uses the currency of species, but remember that biodiversity embraces both finer and coarser levels: genes, populations, and the assembled biomes and ecosystems.

Populations, sometimes genetically distinct, can be in differently dire straits: while lions overall might qualify as Vulnerable, those in West Africa are Critically Endangered whereas those in South Africa are of Least Concern. As for the loss of genetic diversity—which blights Florida panthers and the Kakapo of New Zealand—a modest ambition is that the species has enough genetic adaptability to survive into the future.

Turning to ecosystems and biomes, globally, forests the size of Portugal are felled annually. About half of this area is balanced by regrowing (albeit generally less diverse) forests, so the net annual loss is about 5 million hectares, mostly gone from the tropics. Tropical forests cover less than 10 per cent of Earth's land surface, but support two-thirds of global biodiversity. Their prospects are bleak due to unabated deforestation and the battering of logging, hunting, agricultural expansion, and human settlement. At 3 p.m. on 19 August 2019 São Paulo fell dark, engulfed in smoke from Amazon fires almost 3,000 kilometres away. These fires, caused by deforestation, are likely to continue with catastrophic consequences for biodiversity and carbon emissions. One fateful prediction calculates that the loss and degradation of tropical forest is, irrespective of other anthropogenic stressors, sufficient to precipitate mass extinction.

Marine biodiversity loss increasingly impairs the ocean's capacity to provide food, maintain water quality, and recover from perturbations. Since the 1950s, the catch of coral reef fishes per unit effort has declined by 60 per cent, and global coverage of living coral, and reefs' capacities to provide ecosystem services, have halved. Although only 1.2 per cent of international oceanic waters are protected, an attempt to increase this to 30 per cent under a UN High Seas Treaty failed in August 2022. Plastics were bad enough (90 per cent of albatross chicks on Midway Island had plastic in their gullets), but now microplastics have emerged as a hazard to various aquatic and terrestrial threatened species, communities, and habitats. They provide substrates for harmful bacteria and fungi, and block enzymes that threaten the fertility of humans and much else. The COVID-19 pandemic has led to more than 8 million tonnes of pandemic-associated plastic waste being generated globally, with over 25,000 tonnes entering the global ocean (one glimmer of technological hope comes from the bio-prospecting discovery in 2022 that two enzymes in the saliva of wax moth caterpillars can digest polyethylene plastics).

More locally, in the UK natural wildflower meadows once graced every parish—today only 2 per cent of those that existed in the 1930s remain. With them has gone the shelter and food for pollinators including over 250 species of bee. Bee biodiversity has been further impacted by the sub-lethal effects of neonicotinoids originally intended to improve agricultural yields and enhance gardens. Beyond dispassionate science, all this resonates with Aloysius Horn's evocative warning, 'When man has destroyed nature it will be his turn to go: the barren earth will swallow him up.'

Do extinctions matter?

Extinctions occur naturally—everybody knows the dinosaurs are gone, except insofar as they live on as birds—and what dies out 'naturally' may not be the concern of conservation biologists. Indeed, taking the long view, of the 4 billion species that are estimated to have evolved on Earth over the past 3.5 billion years, 99 per cent are extinct. Further, while we nudge towards a mass extinction, the fossil record shows there have already been five. If extinction is natural, why might biodiversity conservationists think it is their business to intervene?

Outside periods of mass extinction, rates of speciation (the evolutionary process by which populations evolve to become distinct species) and of extinction have been similar. However, nowadays extinction rates are a thousand-fold greater than in the fossil record. Compare the body sizes of all animals that are known to have gone extinct in the Pleistocene (from 2.6 million to 11,500 years ago) with those going extinct in the last 5,000 years. Then as now, extinction was and remains more rampant amongst larger animals. Loosely, the risk of extinction is greatest if geographical range is small, human population density is high, and at high latitude (a proxy for various environmental and anthropogenic factors). However, the correlation between extinction risk and aspects of life history differs between and within taxa. This makes it hard to place a general bet on which risk factors are most deadly; assessing the odds needs local knowledge and, worse, the Millennium Ecosystem Assessment reckons that future extinction rates may increase 10-fold. Indeed, the vortex sucking species down the drain of extinction might be even worse than that, because of disproportionate consequences of the exponentially increasing human population and its accelerating consumption of resources. On top of all that, the mournful conclusion in 2020 of conservationist Gerardo Ceballos and his team was that extinction breeds extinction: the close ecological interactions that bind species mean that as one of those on the brink tumbles over, it tends to drag others toward annihilation. As the human population expands, and its footprint grows heavier, the destruction of nature will inexorably increase unless remarkable technical innovation is supported by radical behaviour change; the latter is the least likely.

Species differ in their roles in ecosystems; elephants engineer landscapes, elephant shrews do not. Moreover some are phylogenetically 'more special' than others: although the aardvark is classified as Least Concern by the IUCN, it is the world's most Evolutionarily Distinct (ED) mammal and the only living representative of an entire order, Tubulidentata. If the aardvark goes extinct, a disproportionate amount of unique evolutionary history would go with it.

Another way in which all species are not equal is in the value that people attribute to them. Not everybody, not even all naturalists, has been inspired by the struggle to save the Bermuda snail, and few, even amongst logical extremists, grieved much over the extinction of smallpox or, amongst much bigger organisms, the parasitic guinea worm. Few 18th-century Britons lamented the extirpation of wolves from Britain, any more, to be candid, than many rural Africans would be sad to see the last of the lions in their backyard.

Oxford conservation researcher Ewan Macdonald led a comparison of which of 100 mammal species different nationalities of people preferred for conservation. For an English-speaking sample from the USA to Australia, the top three species were 1. Tiger (by a huge margin), 2. African elephant, and 3. Lion, and of the top 10, six were big cats. For farmers bordering Hwange National Park in Zimbabwe not a single big cat was in the top 10 (actually several nudged towards the bottom half), and their top three were 1. Zebra, 2. Giraffe, and 3. Scrub hare—all of which you can eat, and none of which eat you or your livestock.

The generality of why extinctions matter is famously expressed by the rivet analogy, with its unsettling reminder that an aeroplane can keep flying as it loses seemingly insignificant rivets until the loss of one rivet too many has catastrophic consequences. So too, the cumulative impacts of species extinctions within an ecosystem may eventually lead to failure of ecological function. The nature of ecosystems is that one thing leads to another, as elegantly illustrated by experimental mimicking of defaunation in Kenya. When fenced exclosures were used to 'remove' species larger than about 15 kg, the consequences included cascading effects on fire intensity, cattle production, disease prevalence, fungal infectivity, photosynthetic rates, and transpiration rates. Exclusion of the big species also changed plant–animal interactions by altering the mutualism between ants and the dominant tree, *Acacia drepanolobium*. This, in turn, drove changes in fruit production, ant defence, herbivory of shoots, thorn production, nectary production, and thorn length. Similarly, the removal of a starfish species from a rocky intertidal community at Mukkaw Bay in the Pacific Northwest allowed mussels to dominate, and they then wiped out many other marine species. A final example: restriction on commercial whaling allowed a rise in killer whale numbers, leading to the loss of sea otters that had previously limited the density of sea urchins which in turn eat kelp. With fewer sea otters, and more urchins, kelp forest densities decreased by a factor of 12. Kelp forests provide food and habitat for a broad range of over 1,000 species of animals and plants, including coastal fish, marine birds, and filter-feeders like mussels and barnacles, all potentially impacted by this intricate cascade. The point is, extinctions reverberate across trophic scales.

Extinctions also threaten natural cycles, processes, and other products that are variously crucial to, or demanded by, humanity. These contributions are called 'natural capital', defined as that part of nature which directly or indirectly underpins value to people, including ecosystems, species, fresh water, soils, minerals, the air, and oceans, as well as natural processes and functions. Natural capital underpins all of the four other types of capital: manufactured capital, financial capital, human capital, and social capital. For example, insect pollination, needed for 75 per cent of food crops, is worth ~10 per cent of the economic value of the world's entire food supply. Pollinators are declining globally, in both abundance and diversity, and with them has gone the abundance of plant species reliant on pollination. Birds and bats are also important pollinators, and in New Zealand decline in pollinating birds has ultimately reduced seed production and plant population regeneration.

Over half (55 per cent) of global GDP, equal to US$41.7 trillion, is dependent on high-functioning biodiversity and ecosystem services. However, a fifth of countries globally are at risk of their ecosystems collapsing due to declining biodiversity and related beneficial services. The undermining of the human enterprise through the unsustainable ravaging of biodiversity and natural resources is neither the prerogative of the Global South, nor is it particularly worse amongst the 40 or so countries populated by the bottom billion of poorest people. Amongst the G20 economies, Australia and South Africa have the most fragile Biodiversity and Ecosystem Services, principally due to water scarcity (predicted to worsen with global warming), and threats spanning pollination services to coastal erosion. Within the G20, Indonesia and Brazil have the highest percentage of intact ecosystems, but not for long: Indonesia, and South-East Asia as a whole, has seen the fastest loss of forest (between 1990 and 2020 the region lost nearly one-sixth of its forests, 376,000 km^2, more than the area of Germany). A deplorable record was set in 2021 of 13,200 km^2 of Amazonian rainforest lost (equivalent to the US state of Connecticut). While mindful of the inequities of history, these and other countries with heavy economic dependency on natural resources urgently need sustainable development policies that protect the biodiversity that is their natural capital and qualifies for respect in its own right.

Causes of extinction

In 1989 Jared Diamond memorably labelled the four main reasons for biodiversity loss as the 'Evil Quartet': (i) overexploitation, (ii) habitat destruction, (iii) introduction of alien species, and (iv) secondary extinctions (e.g. a host-specific parasite). These four can be reshuffled and elaborated, but Diamond was right that the infinite diversity of biodiversity's problems coalesces into a handful of recurring themes, saliently the degradation of habitat (ultimately driven largely by agriculture or forestry), and overuse or persecution, which can be direct, like killing endemic Mauritian fruit bats, or indirect, like accumulating agrochemicals up the food chain.

In the early 1970s, I travelled to London with a group of fellow research students to listen with rapt attention to American ecologist Paul Ehrlich addressing a filled Albert Hall on the perils of the human population time-bomb. Our early generation of conservation biologists widely accepted that the rampant growth in the human population and its deepening footprint was the ultimate cause of the deteriorating state of nature. Indeed 1972 saw the United Nations recognize sustainability and environmental awareness as a global issue through the Stockholm Convention, while leading scientists of the day set out their 'Blueprint for Survival'. When my great-grandfather was born there were fewer than 1 billion people on Earth; since I was born, the tally has grown by a billion people every 12–15 years, with today's population of 8 billion more than double that on the night we listened with foreboding to Ehrlich's warnings. In subsequent decades it became unfashionable to acknowledge, let alone discuss, the fundamental role of human population in the biodiversity crisis. By 2100 the human population is projected to hit 11 billion, and its footprint, which fairness demands must be more equitably spread, will be felt by biodiversity everywhere, however great the alleviating ingenuity of technology. In the early 1970s this was the most pressing subject of the time, and it seems that the pendulum of opinion is starting to swing back to acknowledging this reality, especially in the face of climate change.

What do we need to conserve?

In 2007 in the context of the famous annual migrations of the Serengeti, ecologist Mike Norton-Griffiths provocatively asked: how many wildebeest do you need? That awkward question might be asked of any species, and conservation biologists are likely to answer pragmatically, within the limits of pestilence, as many as we can secure.

Ensuring that no important area of biodiversity is neglected requires a classification scheme that is representative. One global classification is the World Wildlife Fund's (WWF) Ecoregions: 'relatively large units of land or water containing a distinct assemblage of natural communities sharing a large majority of species, dynamics, and environmental conditions'. Ecoregions represent the original distribution of distinct assemblages of species and communities. The WWF recognizes 867 distinct units of terrestrial Ecoregions, classified into 14 different biomes such as forests, grasslands, or deserts. It is important that each biome, and as many ecoregions as possible, are the foci of biodiversity conservation.

The traditional mechanism for protecting biodiversity has been to establish Protected Areas (PAs). PAs have increased markedly since the World Park Congress in 1962: now they cover 15.4 per cent of the world's land area and 3.4 per cent of the global ocean area. Aside from aspiring to prevent the loss of habitat and species, PAs help curb deforestation and, according to the UN's World Conservation Monitoring Centre, support the livelihoods of over 1 billion people. PAs store 15 per cent of the global terrestrial carbon stock. As well as protecting about 3,000 tigers, tiger reserves in India averted forest loss and thus avoided the emission of 1 million tonnes of CO_2. However, PAs fully encompass less than a quarter of areas identified as particularly important for biodiversity, and many terrestrial and marine ecoregions are still poorly represented.

Prioritization

The most painful word in biodiversity conservation is prioritization. How is it to be done? One approach is to consider functional types of species: 'keystone' species—those which have a very high impact on a particular ecosystem relative to their population; 'flagship' species—strategic choices that capture the public imagination; 'indicator' species—representative of the state of other species; and 'umbrella' species—those that demand ecosystem requirements that also safeguard other species.

An important notion is complementarity—colloquially, maximizing bangs per buck. For example, how best to prioritize where to conserve the 47 species of carnivore found in North and Central America? One approach is to plot their distributions to reveal spatial patterns in their overall richness, regions where they are endemic, and threatened status, and thereby to identify the smallest number of map squares whose protection would safeguard all 47 species.

Countries differ in the degree to which they prioritize conservation of biodiversity. Megafauna are particularly valuable in economic, ecological, and societal terms, and are challenging and expensive to conserve. The Megafauna Conservation Index assessed the spatial, ecological, and financial contributions of 152 nations towards conservation of the world's terrestrial megafauna. Although 90 per cent of countries in North/Central America were major or above-average performers, so too were 70 per cent of generally poor countries in Africa, while approximately one-quarter of generally rich countries in Europe and Asia were disappointing under-contributors.

Countries with the lowest GDP tend to sustain most biodiversity, often along the Equator, where political issues can take priority over conservation. At least part of the cost of conserving biodiversity, therefore, needs to be borne by the global community. This need for cost-sharing emphasizes that while conservation may be underpinned by science it is operated by politics. There is problematic bias where conservation work is done, especially biodiversity surveys. Some ecosystems, taxa, and locations (especially in sub-Saharan Africa and South America) are under-represented, probably because they are inaccessible to researchers and uncomfortable for agencies, notably due to armed conflict.

Biodiversity conservation involves tough choices that go beyond the nuances of biological priorities to realpolitik. One of the first studies to combine these aspects was led by Oxford WildCRU's Amy Dickman, whose team explored two dimensions to prioritizing the conservation of 36 species of the cat family, Felidae. They combined biological factors (endangerment, evolutionary distinctiveness, body mass, percentage of range protected, habitat, and range overlap) with metrics of economic and political instability. Some species, such as the Iberian lynx, live in privileged surroundings, but 62 per cent of the countries where tigers occur face serious infrastructural difficulties and poor governance—these disadvantaged countries comprise 80 per cent of the tigers' range. The point is that conservation approaches should be tailored for each country and circumstance.

What's to be done?

The United Nations' 2015 Sustainable Development Goals (SDGs) require reversing trends in biodiversity, due to habitat loss and degradation, while feeding the global human population. One team of modellers explored seven scenarios that might reverse biodiversity trends by varying supply side (sustainably increased crop yields and increased trade of agricultural goods); demand side (reduced waste of agricultural goods from field to fork and diet shift away from animal calories); and conservation efforts (increased extent and management of protected areas and increased restoration and landscape-level conservation planning). They concluded that the SDGs could be achieved in theory, albeit with unprecedented global coordination. But in practice an increase in the extent of land under conservation management, restoration of degraded land, and generalized landscape-level conservation planning represents a huge political agenda. Nonetheless, the models suggest that with political will positive trends in biodiversity could be delivered by the mid-21st century. Furthermore, with greater emphasis on sustainability in trade (an aspiration directly addressed at the 2021 second United Nations Global Conference on Sustainable Transport), reduced food waste and more plant-based human diets, the models suggest that over two-thirds of anticipated biodiversity losses could be averted and biodiversity's decline due to habitat conversion reversed by 2050. Other major threats, such as climate change and biological invasions, can predominate locally, and are likely to worsen, and interact damagingly with land-use change.

Whatever the theory, the point is that aligning this panoply of interacting factors poses a daunting political challenge. Against this genre of hopeful grand plans, the aphorism that the road to hell is paved with good intentions is painfully topical in the failure to achieve, for the second decade in a row and despite many improvements, any of the 20 Aichi biodiversity targets agreed in Japan in 2010 to slow the loss of global biodiversity and destruction of ecosystems.

Protecting the planet's biodiversity can be achieved directly or indirectly and at scales from local to global. Although irresponsible large-scale fishing and mining, the actions of some multinationals, and the disproportionate impacts of wealthy lifestyles adversely affect the environment, many impacts on biodiversity and the wider environment are ultimately not the dirty work of avaricious evildoers, but are commissioned by people like you and me, and so it is at least partly with you and me that the buck stops regarding behaviour change as consumers, and encouraging the political change that guides legislators. Sometimes that change is direct—persuading or compelling somebody not to consume, hunt, keep as a pet, or degrade the habitat of a particular species. Sometimes it is indirect—persuading a city-dweller not to demand cheap food from the far-off countryside, or an even further-off part of the world. Whether the impacts, and necessary remedies, are direct or indirect, local or global, humanity lives in a bounded system within which natural resources, living and inanimate, are limited and limitedly renewable. The self-interest of sustainability ultimately dictates that natural capital should be consumed with care—according to the precautionary principle—and with respect.

E. O. Wilson, the visionary Harvard biologist, concluded controversially that the human enterprise should, and could, be reconfigured so that half the earth was dedicated to protecting biodiversity. The Half-Earth Project reignited the long-running, but nonetheless crucial, debate on whether people and farming should coexist with, or apart from, nature: land-sharing versus land-sparing. In practice, this contrasts the food security and biodiversity consequences of local farming practices that variously integrate or separate food production and biodiversity, along a scale from intensive, mixed, organic, regenerative, to wilded. How to optimize the outcomes lures modellers into variously eccentric flights of fancy. One simulation, at the extreme of separating agriculture and biodiversity, involved global optimization of fertilizer-use and 16 major crops being strategically allocated across the world's croplands. The simulated outcome happily plugged current gaps in crop yields, reduced land-take by half, but banished all save the most skeletal biodiversity from cropland. The 'spared' land would be associated with decreased greenhouse gas emissions and irrigation requirements, and would facilitate carbon sequestration in restored natural vegetation. But what political system might deliver this outcome?

A different modelling exercise, called Ten Years For Agro-ecology, explored the sharing perspective. The simulated goal was to provide sustainable food for Europe's 550 million citizens, whilst preserving biodiversity and natural resources and limiting climate change. These simulations convinced the researchers to advocate an urgent transition to an agro-ecological system based on the phasing-out of pesticides and synthetic fertilizers, widespread adoption of agro-ecology, phasing-out of vegetable protein imports, and the adoption of healthier diets by 2050. All this would hypothetically deliver healthy food for Europeans while reducing Europe's global food footprint and cause a 40 per cent reduction in greenhouse gas emissions from the agricultural sector while regaining lost biodiversity.

The down-to-earth reader might be sceptical of models predicated on the utopian prospect of the world's nations agreeing to allocate croplands on an optimal global scale. But even having the discussion flushes out the extent to which the human enterprise depends inescapably upon natural ecosystems, where the crucial currency is Biodiversity and Ecosystem Services, and where the increasingly dire state of biodiversity may compel levels of cooperation and intergenerational pacts without precedent. My own futuristic hope, already within technological reach but requiring scaling, lies in freeing land from agriculture by feeding people on microbial proteins and fats and other high yield, low impact, forms of food production. The topic of biodiversity conservation might at first glance have been mistaken as the preoccupation of natural historians; instead, it turns out to be the foundation of humanity's future.

Chapter 3
What is the purpose of biodiversity conservation?

Because it is a practical discipline, the question, 'what is the purpose of biodiversity conservation?' is entangled with the applied question of 'what do conservationists do?' A glib answer could be to document, understand, and remedy the loss of biodiversity described in Chapter 2. Going deeper, Aldo Leopold thought in 1949 that it was for 'preserv[ing] the integrity, stability and beauty of the biotic community'. Charles Elton, often considered the father of ecology, wrote in his 1958 book *The Ecology of Invasions by Animals and Plants*, that conservation was important: (1) to promote ecological stability; (2) to provide a richer life experience; and (3) because it is the right relation between humans and other living things. His points progress from science to ethics and he expanded the third point, saying, '...there are some millions of people in the world who think that animals have a right to exist and be left alone, or at any rate they should not be persecuted and made extinct as a species'.

In his 2022 textbook *Effective Conservation* Ignacio Jimenez summarizes, 'Our task is to maintain and/or restore natural ecosystems with most of their original components, thereby generating maximum benefits for and support from society.' Hambler and Canney's 2013 textbook offers, pithily, 'conservation is the protection of wildlife from irreversible harm'.

In the past, conservationists discussed whether they should focus on particular species, generally ones that are in trouble, or on the habitats that provide a living for those species, threatened or otherwise. Both foci aspire to efficiency, insofar as the conserved species serves as an umbrella, even an ambassador, for other species, while the conserved habitat provides a living for a whole community. More recently, some conservationists have tended, instead, towards conserving ecological processes, such as the top-down influences of predation, which at an extreme of rewilding aspires to reconstitute broken processes even with the introduction of quite different, but analogous, actors on the stage.

In 2001 Tony Whitten pointedly titled a paper 'Conservation biology: a displacement behavior for academia?' It is tempting, when considering purpose, to be deflected towards the question of what conservationists do, and indeed there may be clues to the former in the latter. Sometimes the answer is reminiscent of gardening, such as efforts to maintain habitats for species such as pearl bordered fritillary or corncrake that, a bit like steam engines, no longer quite fit with modern human land-uses. Sometimes it seems more like restitution, putting right, brutally, if with a heavy heart, the wrongs of an earlier generation, such as killing the beautifully adapted but invasive American mink where it harms native species. Sometimes conservation takes the form of a sort of community engineering seeking to maintain a particular balance of predators and prey or competitors. Or, somewhat like UN peacekeepers, brokering reconciliation between people and wildlife in conflict. Or, at a much bigger scale, choreographing land-uses for example with a goal to keep protected areas in good heart, to maintain ecosystem functionality and thus services, or, further beyond biology, to understand and nudge the national and geopolitical budgets that frame such activities. These activities and interventions are the worthy work of skilled technicians deploying the analogues of scalpels, prophylactics, antibiotics, and, perhaps especially, the analgesics of palliative care in the doctor's medical bag. By analogy, this is what physicians do, but it is not the purpose of physicians. For that we must turn to the Hippocratic Oath or, were they lawyers, to Lady Justice who weighs the evidence blindfolded. In later chapters I will describe practical problems besetting biodiversity, and the tools used to remedy them, but first I ask: what higher principles direct the actions of conservationists in parallel to those lofty principles of medicine and jurisprudence?

This question, surprisingly, has been little discussed amongst conservation biologists, but it's fair to say that two emphases exist, the first more deeply rooted in history, the second highly contemporary. Michael Soule's foundational essay of 1985 recognized both functional (e.g. a set of fundamental axioms, derived from ecology, biogeography, and population genetics) and normative (e.g. biotic diversity has intrinsic value, irrespective of its instrumental or utilitarian value) tenets of conservation. Twenty-five years later, Peter Kareiva and Michelle Marvier framed, with robust pragmatism, an emerging modern conservation as deeply anthropocentric: to be both defensible and feasible biodiversity conservation must deliver benefits to people. Their concluding sentence, 'In summary, we are advocating conservation for people rather than from people', was eye-catchingly strident. Perhaps that stridency underlay the emerging dichotomy between two types of biodiversity conservationist: the anthropocentric New Conservation scientists and the classical advocates of nature. Woven into these, still surprisingly infrequent, discussions is the contrast between those conservationists who, looking at the same landscape, seek to 'exploit as much as desired without infringing on future ability to exploit as much as desired' or 'exploit as little as necessary to maintain a meaningful life', as incisively characterized by John Vucetich and Michael Nelson in 2010. Soule's original mention of intrinsic value will turn out to be pivotal to where individual conservationists sit on this spectrum.

Not only has the purpose of conservation been changing, so too has its ambit and remit. In 2007, Macdonald, Collins, and Wrangham captured conservation's journey amongst disciplines: 'Environmental problems can no longer be solved by the traditional algebra that isolates issues one at a time—but must be treated as an ensemble that is addressed as a whole. Getting conservation right requires developing practical solutions to the most complicated simultaneous equation ever written!'

This central shift in perspective is illustrated by the history of England's statutory conservation agency. In 1947, Command 7122 of the then Ministry of Town and Country Planning created England's Nature Conservancy. In 2005, several entities later, Natural England was created to fulfil the former Nature Conservancy's role as the statutory body for English nature conservation. In 1947, in an era when conservation was the prerogative of enthusiastic natural historians thinking along lines now referred to as 'fortress conservation', five functions were ascribed to protected areas of which the last, and most briefly mentioned, was 'amenity' (i.e. concerned with benefits for people as opposed to safeguarding nature, largely from people). At its launch in 2005 the strap-line, which I had a hand in drafting, for Natural England was 'for people, places and nature'—people conspicuously listed as first of these beneficiaries. (And it continues with a route map to delivery that 'put people at the heart of our work', in a statement of strategic direction richly peppered with words like 'enable', 'empower', 'share', 'deliver', and 'engage'.) Interestingly, in the context of questing to discover the purpose of conservation, Natural England included in the justification of its mission the 'intrinsic value of wildlife'. Natural England published its 'Strategic Direction' in 2014 and, like Soule, mentions 'intrinsic value'—twice, in fact: 'Ultimately any approach to conservation will only be successful when it starts from a real public consensus on the value of the natural environment—*for both its intrinsic value* and its value to people and the economy', and the closing tag-line is: 'Natural England is here to conserve and enhance the natural environment, *for its intrinsic value*, the wellbeing and enjoyment of people and the economic prosperity it brings' (my italics). As mentioned, who or what has intrinsic value will turn out to be important. Interestingly, the missions of the US Fish and Wildlife Service (to protect species in need and pursue their recovery) and the US Environmental Protection Agency (to protect human health and the environment) say nothing of intrinsic value.

Conservationists write much more about their practice than they do about any higher principle that guides them. However, because modern biodiversity conservation is so deeply holistic, and permeates the entire human enterprise, it is important to think about those principles. For example, conservation abuts social justice when humans and wildlife are in conflict—an intersection which led John Vucetich and colleagues in 2018 to propose two principles:

> Humans should not infringe on the well-being of others (including other humans, large carnivores, or other parts of nature with intrinsic value) any more than is necessary for a healthy, meaningful life. When the ability to live a healthy, meaningful life genuinely seems to infringe on the wellbeing of some intrinsically valuable element of nature (such as large carnivores), then the just solution will less often be found in depriving large carnivores and more often be found in rectifying an unjust inequality among humans.

This latter principle also focuses attention on ultimate causes of species loss and conflict. These deep-seated drivers of biodiversity loss seem a long way from the naturalistic idyll that motivated many conservationists. They include gross inequalities in wealth distribution within and among nations, and in the costs experienced as a result of conservation, and the pervasive influence of plutocracy. Behind almost every conservation dilemma lie consequences of human population growth.

What higher principles might guide conservationists? One, about which conservation scientists (as distinct from lobbyists) care deeply, like doctors and lawyers, is the primacy of evidence and the rigour of impartiality when assessing it. You might think that reverence for evidence is principle enough, but doctors and lawyers go further. Just as life and justice are principles of medicine and jurisprudence, two higher principles are candidates for guiding biodiversity conservationists: first, safeguarding the intergenerational equity of biodiversity, which can be expressed in terms of protecting natural capital; second, respecting the intrinsic value of non-human beings.

Safeguarding intergenerational equity

In Chapter 2 I described how our predecessors, and indeed you and I, have, since the industrial and agricultural revolutions, already widely, and in many cases irreparably, degraded the biodiversity that humanity can benefit from. To continue this trend will be an infraction of the principle of doing no harm. Joseph Stiglitz, who won the Nobel Prize for Economics in 2001, was among the first modern thinkers to address rigorously the concept of this intergenerational fairness or equity. The idea has its roots with the 18th-century philosopher Edmund Burke, who argued that 'society is but a contract between the dead, the living and those yet to be born'. He argued forcefully that no generation should be so selfish as to think only of themselves. Rather, and with a sentiment echoed in the 21st-century Extinction Rebellion and Climate Action movements, Burke saw people as guardians with an obligation to sustain the natural environment in order to bequeath it to following generations. This raises the question of the discount rate for future generations against the needs of the present generation.

In his inspirational work *Green and Prosperous Land*, Dieter Helm summarizes the position: 'to deprive future generations of an enhanced natural environment would be to fail in our duties to them, and fail in our specific duty as stewards of natural capital'. Helm is no fringe figure; he is an incisive economist and a UK government adviser, and note particularly his use of the word enhanced. He explains this forthrightness by arguing that the next generation should get an enhanced environment because the current one—thanks to the 'great acceleration' wrought by the Anthropocene—is already not sustainable. Importantly, Helm concludes his economic argument with a moral one: 'We owe it to them to make good on the damage we have done, as well as to make sure it does not get any worse.'

Arguments that almost all of biodiversity contributes to natural capital and thus has utilitarian value are compelling. However I would fear for much biodiversity if the arbiter of its value was only money—quite a lot of rivets could be lost from the metaphorical aeroplane without making any plausible financial difference. If biodiversity is to be saved, the motivation for doing so must include, but extend beyond, monetizable value.

Intrinsic value of non-human species

A second ethical principle for conservation might be phrased in terms of the intrinsic value of non-human beings. But what has intrinsic value? Clearly followers of Soule's type of conservation think at least some wild species have it (and Natural England may think even nature as a whole has it). Objects with intrinsic value are valuable for their own sake, without regard for their utility. Something might be thought to have intrinsic value if, like you, it has interests; and if something possesses intrinsic value the ethical argument is, essentially, that you have an obligation to treat it fairly or with respect and with at least some concern for its well-being or interests.

If you have a dog, or even a hamster, my guess is that you recognize that it has interests, for example in avoiding pain, staying alive, tending its pups, chasing its ball. If you consider your dog has interests you would find it hard to argue that wolves don't, or foxes (having lived with both dogs and foxes I can tell you that foxes think quicker), or indeed any other mammal, and perhaps any vertebrate. Give some thought to where to draw the line, but if you think intelligence is relevant, don't forget octopuses. Anyway, the philosophical point is that possessing intrinsic value entitles an individual to fair treatment with at least some concern for its well-being. This is the view I hold, and it is called non-anthropocentrism (not to be confused with misanthropy). It is the belief that at least some portions of the non-human world possess intrinsic value, and that it is wrong to infringe on the well-being of an intrinsically valuable agent without an adequate reason for doing so. Understanding what counts as an adequate reason would be at the heart of a higher principle guiding biodiversity conservation. How would such a principle—in the context of human–wildlife conflict it can be called Just Conservation—affect your behaviour? With regard to biodiversity conservation the answer can be informed by a thought experiment developed by John Rawls in 1971, and known as the veil of ignorance (it has much in common with Adam Smith's 1759 'impartial spectator'). It boils down, colloquially, to putting yourself in the other being's shoes, and then doing, as far as possible, as you would be done by.

Intergenerational equity, intrinsic value, and respectful engagement

People have been writing about the right way to engage with non-human animals for a couple of thousand years (and depictions in the Maros Caves suggest they've been thinking about it for 35,000 years or so), so it's not easily solved. Nonetheless, attention to intergenerational equity and intrinsic value, roughly characterized as fairness to future generations and to other species, does offer ethical guidance to down-to-earth biodiversity conservation. In an essay I wrote with Fran Tattersall in 2001, I characterized the attitude I favour towards biodiversity as *respectful engagement*. This position built on the observation that not only are the problems people encounter with wild animals complicated and various, prompting reactions from irritation to despair amongst people who may, almost simultaneously, display tireless compassion and deadly loathing. Human–wildlife relationships are nuanced: the fact that habitual lamb killers are rare amongst foxes does not mean that no foxes are habitual lamb killers. Although nature should be cherished for both scientific and moral reasons, the fact that most foxes don't kill lambs means that the shepherd should not act fruitlessly against those that don't, but does not mean he should not act against the one that does.

Nowadays, in many developed countries, the reality of living with wildlife is increasingly distant and both the thrill and the anguish of that reality may be muted. Nonetheless, amongst those who experience nature directly there are some who reject the extreme of brutal utilitarianism but nonetheless do not consider every life sacred. Those taking that position can embrace a fervent desire to conserve and foster with an acceptance of a need sometimes to manage and to use, but for each of those engagements to be respectful. For those for whom the notion of respect is too vague, it might be characterized as taking the responsibility to be aware of all aspects of the issue. That is indeed the holism that transdisciplinary conservation strives to capture.

The idea of respectful engagement with nature raises the thought that the purpose of conservation is somehow to foster coexistence. The injunction never knowingly to act, or support action, that worsens coexistence, and to strive to improve the well-being of wildlife, leads to a general principle: a conservationist has a responsibility to avoid any detriment to coexistence, to foster better coexistence between humans and wildlife, and to improve the well-being of wildlife. The consequences of that principle for a conservationist's judgements and actions will depend on the inclusiveness of his or her moral family, that is, whether an anthropocentric or non-anthropocentric stance is adopted. In the former case, where the circle of species acknowledged to have interests and thus intrinsic value encompasses only humans, then any practice that does not worsen current coexistence and the well-being of wildlife could be acceptable, and the prevailing ethic would be solely utilitarian; an intervention would be appropriate so long as its consequences did not infringe on peoples' future ability to exploit wildlife as much as desired (although remember Dieter Helm's enhancement point). Alternatively, a conservationist adopting a non-anthropocentric stance would by definition believe that the circle of species with interests, and thus with intrinsic value and deserving consideration, would be wider. From this standpoint, the non-anthropocentrist conservationist's responsibility would be (like his or her anthropocentrist colleague) to censure actions likely to be detrimental to coexistence and to strive for outcomes that protect species, and the processes that connect them. However, in addition a non-anthropocentrist would consider also the point of view of the wildlife, which is likely to push judgements in the direction of reluctance to damage the interests of wildlife unless an intervention is required to ensure the ability of people involved to have a meaningful life. The distinction—to include or not consideration of the point of view of wildlife—may seem a small one, but it may have significant impact on the disposition of the two practitioners. For instance, an anthropocentrist conservationist considering the factors affecting the appropriateness of trophy hunting lions should consider a wide diversity of factors, but the imagined viewpoint of the lion is not amongst them. The non-anthropocentrist conservationist evaluating the same topic should consider all the same factors, and in addition the viewpoint of the lion; this may not change the judgement, but it will change the perspective on making that judgement. It is also not straightforward to guess what assessment of pros and cons you would make if you were the lion.

How might a conservationist decide which perspective he or she should adopt? The similarities between mammals, for example, demonstrate they too possess interests just like humans. So we might expect that among conservationists the second stance would prevail, extending considerably beyond humans the species afforded concern. This is not an unfamiliar perspective in many cultures. A gentle and eloquent example comes in botanist Robin Wall Kimmerer's account of Native American attitudes to coexistence with nature, *Braiding Sweetgrass*. Similar thinking is familiar amongst traditional naturalists, as beautifully expressed by Henry Beston in *The Outermost House:*

> We need another and a wiser and perhaps more mystical concept of animals.... In a world older and more complete than ours they move finished and complete, gifted with extensions of the senses we have lost or never attained, living by voices we shall never hear. They are not brethren, they are not underlings; they are other nations, caught with ourselves in the net of life and time, fellow prisoners of the splendour and travail of the earth.

Yet there is variation in values even amongst conservation professionals, a sample of whom were recently found to have attitudes commensurate with one of four distinct purposes for conservation: (i) new conservation emphasizes averting the biodiversity crisis primarily for the ultimate purpose of benefiting human well-being; (ii) orthodox conservation emphasizes that species and ecosystems possess intrinsic value; (iii) compassionate conservation may be seen as a reaction to the increasing use of killing as a conservation tool; and (iv) maintaining and protecting hunting heritage.

The journey to transdisciplinary conservation

With these thoughts in mind, I return to a definition of conservation I helped develop from a non-anthropocentric perspective:

> Conservation—maintaining and restoring the health of ecological collectives—namely, species and native populations and ecosystems.

> Conservation is a constituent element of sustainability. Sustainability—meeting human interests in a socially-just manner without depriving species, native ecosystems or native wildlife populations of their health.

This raises the question of whether conservation and wildlife management (even pest control) are different. At a superficial level they clearly can be, insofar as management and control sometimes involve killing individuals of a problematic species, and deliberately reducing their numbers. However, assuming the 'adequate reason' mentioned above, the same knowledge base (such as the ecological basis of population dynamics) and the same suggested ethical frameworks, and indeed the same attention to animal welfare, should all be applicable (an aspect made explicit in the Compassionate Conservation movement). My own view, therefore, is that the conservation of the imperilled and the control of the pestilential are all embraced by the same framework of scholarship and ethics, and are thus simply different parts of the wide spectrum of challenges to coexistence between people and wider biodiversity.

With this broad remit, what will biodiversity conservation look like in future? Its modern roots began with the social geographers of the late 19th century, along with such writings as those by Aldo Leopold and John Muir. But it leapt forward with the explosive growth of behavioural ecology in the 1960s–1970s, which established biodiversity conservation as a science with its centre of gravity in university Departments of Biology. Indeed, my own research group, the WildCRU at Oxford, grew out of the creation, in 1986, of the first ever university-based research post explicitly dedicated to wildlife conservation. Then came the age of Geographic Information Systems (GIS) and the generation of conservation research launched from Departments of Geography. Soon, other subjects, beyond the natural sciences, joined the interdisciplinary mix with recognized subdisciplines in conservation economics, conservation law, conservation marketing, conservation ethics, and conservation geopolitics. Now business schools too are engaging; while the value of non-human animals greatly transcends that which is monetizable, their financial value will heavily determine their destiny. Realizing that value will require the cleverest business minds (guided by the restraining hand of ethicists). I think the future lies in transdisciplinarity which works across disciplinary boundaries to generate holistic forms of knowledge.

The goal of transdisciplinary conservation is to integrate with organismic and environmental sciences the assemblage of higher-level insights offered to conservation by economics, psychology, political science, law, sociology, international relations, development, ethics, and disciplines with less quantitative epistemologies such as anthropology, environmental history, and human geography. These disciplines together inform choices, and effect behavioural change, at scales from individuals to empires. Such change is most vibrant at the potent intersection of top-down (nations) and bottom-up (citizens) approaches. Transdisciplinary biodiversity conservation should be a major subject on the world stage, central to scholarship and to society, because biodiversity is a crucial working part of the environment on which humanity depends and which, in addition, fuels an aesthetic purpose which might be called spirituality. Furthermore, the discussions and policies that this transdisciplinary conservation must permeate spans an ever-wider stage. In addition to wrestling the particularities of endangered Bermudan snails, the Lord Howe Island stick insect (the world's rarest insect, Figure 8), the orchid *Dracula mendozae* (Figure 9, known from only one specimen), pink pigeons, and Ethiopian wolves, it must simultaneously embrace challenges as immense as Net Zero carbon accounting, debt-for-nature swaps, and the quest for nature-based solutions. And it must do so with care for the well-being of both biodiversity and people.

8. Lord Howe Island stick insect or tree lobster.

9. Orchid *Dracula mendozae*.

What might this critically important discipline aspire to deliver? In an essay published in 2013 I suggested that one answer might be 'a human population enjoying a healthy, equitably high and sustainable standard of living, alongside functioning ecosystems populated with "natural" levels of biodiversity'. Delivering such a vision requires almost unimaginable ingenuity, rationality, and political acumen alongside valuing the inspirational, even a feeling of the spiritual, in nature. It will also require robust frameworks of regulation, enforcement, and law, and so I close this chapter with a brief introduction to some of the important existing supra-national agreements.

The rules of the game

The Convention on Biological Diversity (Biodiversity Convention or CBD) was adopted at the Conference on Environment and Development (the 'Earth Summit') in Rio de Janeiro, Brazil in June 1992, and entered into force in December 1993; it has been ratified by 196 nations and covers biodiversity at all levels: ecosystems, species, and genetic resources. As the first global treaty to provide a legal framework for biodiversity conservation, the Convention established three main goals: (i) the conservation of biological diversity; (ii) the sustainable use of its components; and (iii) the fair and equitable sharing of the benefits arising from the use of genetic resources.

Some commentators think that the outcomes of the 15th United Nations Biodiversity Conference of the Parties (COP15) to the UN Convention on Biological Diversity (CBD), notably the Montreal–Kunming Global Biodiversity Framework (GBF), was a historic watershed in the seriousness with which governments globally treat conservation. These optimists note that on 19 December 2022 the 196 countries that are Parties to the convention (which notably excludes the United States) approved an agreement which includes conserving 30 per cent of land and oceans by 2030 ('30 by 30'; the 2010 Aichi target had been

17 per cent of land and 10 per cent of oceans) and 22 other targets intended to reduce biodiversity loss, including an agreement to reform $500 billion of subsidies that are harmful to nature, and to increase biodiversity financing to developing countries. Brian O'Donnell, director of the Campaign for Nature, said, 'The "30 × 30" target marks the largest land and ocean conservation commitment in history.' Target 15 importantly requires companies and financial intitutions, especially large and transnational ones, regularly to monitor, assess, and transparently disclose their risks, dependencies, and impacts on biodiversity. Nonetheless, pessimists observed that the agreement is not legally binding and lacked quantifiable pledges around reducing the production and consumption that drive biodiversity loss. They remember that, despite considerable progress (<https://www.cbd.int/aichi-targets/>), the world failed to meet in full a single one of the 20 Aichi biodiversity targets set more than a decade previously in Japan.

The GBF groups its 23 targets into four goals to be achieved by 2030. (a) The integrity, connectivity, and resilience of all ecosystems are maintained, enhanced, or restored by 2050; human-induced extinction of known threatened species is halted, and, by 2050, extinction rate and risk of all species are reduced 10-fold; the genetic diversity within populations of wild and domesticated species is maintained. (b) Biodiversity is sustainably used and managed and nature's contributions to people, including ecosystem functions and services, are valued, maintained, and enhanced by 2050. (c) The monetary and non-monetary benefits from the utilization of genetic resources are shared equitably, including with indigenous peoples and local communities, and substantially increased by 2050, thereby contributing to the conservation and sustainable use of biodiversity, in accordance with internationally agreed access. (d) Adequate means of implementation, including financial resources, capacity-building, technical and scientific cooperation, and access to and transfer of technology to fully implement the GBF are secured for all Parties, especially developing countries, progressively closing the biodiversity finance gap of $700 billion per year, and aligning financial flows with the GBF and the 2050 Vision for Biodiversity.

In the realm of policy, words matter. Seasoned conservation economist Frank Vorhies noticed that the root 'wild' (wildlife, wilderness) was used more than 90 times in 1980 in the *World Conservation Strategy: Living Resource Conservation for Sustainable Development*, by IUCN, UNEP, and WWF with FAO and UNESCO. As years and CBD COPs passed, this word became endangered and was effectively extinct (one mention) in the 2010 CBD COP10's new *Strategic Plan for Biodiversity and Aichi Targets*. However, in the 2022 GBF it reappears seven times: a case of rewilding that may cheer wildlife conservationists, while also prompting thought on how best to define 'wild', and distinguish it from non-wild.

As an environmental treaty of the United Nations, the Convention on the Conservation of Migratory Species of Wild Animals (CMS) provides a global platform for the conservation and sustainable use of migratory animals and their habitats. CMS brings together the states through which migratory animals pass, the 'Range States', and lays the legal foundation for internationally coordinated conservation measures throughout a migratory range—it is the only global convention specializing in the conservation of migratory species, their habitats, and migration routes.

IUCN, the International Union for Conservation of Nature, was established on 5 October 1948 in the French town of Fontainebleau. As the first global environmental union, its aim was to encourage international cooperation and provide scientific knowledge and tools to guide conservation action. Today, with the expertise and reach of its more than 1,300 members—including states, government agencies, NGOs, and Indigenous Peoples' Organizations—and over 15,000 international experts, IUCN is the world's largest and most diverse environmental network.

The Convention on International Trade in Endangered Species of Wild Fauna and Flora (CITES) is a multilateral treaty to protect endangered plants and animals. It was drafted as a result of a resolution adopted in 1963 at a meeting of members of the IUCN. Its aim is to ensure that international trade in specimens of wild animals and plants does not threaten the survival of the species in the wild, and it accords varying degrees of protection to more than 35,000 species of animals and plants.

These various treaties often operate in a far from optimal manner. For example hunting trophy export quotas established for African range states for leopard under CITES were arbitrary, lacking robust scientific bases, without regular adjustment, fundamentally at odds with the principles of sustainable use, precaution, and adaptive management. Nonetheless legal instruments and experts can benefit wildlife conservation in the 21st century through transdisciplinary research, ethically informed and actively applied.

Part 2

The Big Five

Chapter 4
Invasive species

A grey squirrel's acrobatics in park or garden may be as close to the delight of wild nature as many British city-dwellers get. It is also a poignant example of the collision between values provoked globally by invasive species. As its scientific surname, *Sciurus carolinensis*, betrays, the grey squirrel did not originate in England; it was introduced deliberately, and thereafter repeatedly, from America in the 1870s, to rural estates whose owners thought enriching their fauna was the right thing to do. Subsequently it has spread throughout mainland Britain, almost entirely displacing the native red squirrel. Nowadays, the loathing with which some regard the grey squirrel is captured by jingoistic reference to 'American tree rats', while others watch the bushy-tailed visitor captivated from suburban windows; either way, the public is broadly aware that non-native species are often problematic. Though some ethicists might struggle to see why it is meritorious to poison, trap, or shoot grey squirrels but criminal to do the same to red ones, most people agree that, with hindsight, replacing a native pest of forestry with a more damaging non-native one was a mistake. The biology of invasions is technically complicated. But this example highlights the complexity of wider issues that permeate invasion biology, entrenched as it is in economic and ethical challenges that thwart consistency and guarantee controversy. Being in the wrong place at the wrong time is often awkward for all concerned.

Consider Antarctica. For 15 million years, all the organisms that comprise the working parts of the Antarctic ecosystem have evolved untroubled by the outside world. But in 2022 the British Antarctic Survey documented ships, largely carrying tourists, their hulls alive with mussels, barnacles, crabs, and algae, arriving there from 1,500 ports worldwide. This bombardment (currently involving about 70,000 tourists yearly) will surely accelerate the variant of ecological dysfunction known as biofouling.

Even if most invasive species do not devastate native biodiversity, they do generally change things, often in ways that conservationists perceive as unwelcome. In the UK, 5 per cent of all priority species with Biodiversity Action Plans (9 per cent for vertebrates) and 32 per cent of priority habitats list non-natives as a threat. Scouring the IUCN Red List for the causes of extinction of 247 species of plants, amphibia, reptiles, birds, and mammals that disappeared since AD 1500, aliens emerged as the second most important driver to doom overall. For vertebrates they are the most important. Aliens comprised 48 per cent of all the threats that had blighted the extinct mammals and 68 per cent for all the island endemics. Around 30 taxa of aliens had been involved in these 247 extinctions; for extinct mammals, feral cats and rats predominated as the villains—representatives of ferals (domestic animals that have lapsed into a wild form) and human commensals (species that live in close association with humans).

Currently, species are being redistributed between continents as they leak from unprecedented global trade. In a sinister game of biogeographical musical chairs the movements of species are further encouraged by man-made climate change. In Yellowstone, an invasive beetle advantaged by climate change threatens to devastate native conifers. In China there was a 10-fold increase in the number of invasive species between 1990 and 2005. In the United States, 35 per cent of vertebrate families in which there is live trade are now established as introductions beyond their native range. Natural community resilience deteriorates under interacting pressures from habitat loss, hunting, and climate change—opening wider a door to invaders that further degrade ecosystems. Under Article 8(h) of the CBD, signatories are required to prevent the introduction of alien species, and to control or eradicate those alien species which threaten ecosystems, habitats, or species.

The ecological costs of invasives may be incalculable, but what of the financial costs? These include the costs of damage, and of preventing worse damage, once the invasive has taken hold. Remembering the adage in Public Health that prevention is better than cure, these costs should be compared to those of keeping invasives out in the first place. There are 10,822 species listed on the European Invasive Species Gateway and about 10 per cent of those have significant ecological impact, which in 2008 was estimated to cost the EU at least 12 billion Euros annually. From 1970 to 2017 invasives globally cost at least $1.288 trillion (at 2017 US dollar rates) (Figure 10), perhaps up to 30 times more. The annual mean cost was $26.8 billion (for perspective, greater than the annual GDP of 50 of 54 African countries). But this shrouded a doubling every six years, so the annual cost in 2017 was $162.7 billion. The costs of damage increased at twice the rate of investment in its management.

Invertebrate invasives cost more than vertebrates, which cost more than plants. Overall in 2017 invertebrate invasives cost $23.8 billion, vertebrates (of which 88 per cent were mammals) cost $1.3 billion (Figure 11). Free-roaming, unowned, feral cats annually kill 12.3 billion mammals and 2.4 billion birds in the USA alone, where feral pigeons are estimated to cost $1.1 billion dollars p.a. Damage from dingoes (considered as native since 1992) and other wild dogs in Australia are reported at A$48.5 million p.a.; rabbits, A$206 million p.a. Although there is no comprehensive, harmonized, and robust compilation of the costs of biological invasions worldwide, we can be confident it is a lot.

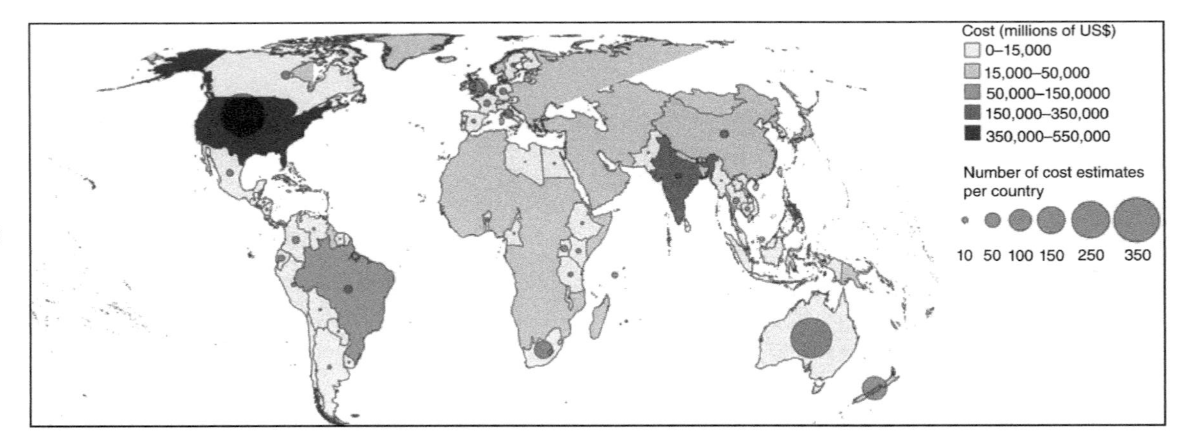

10. Geographical distribution of the costs due to alien species for 1970–2017.

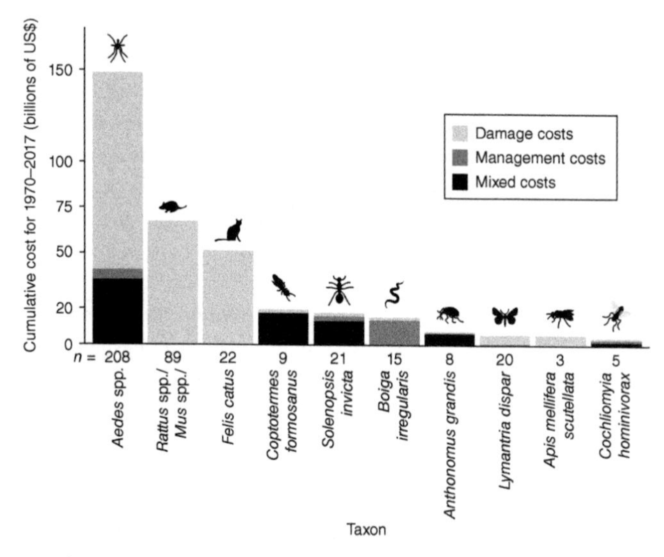

11. The 10 costliest taxa for cumulative damage and management costs (2017 prices) 1970–2017. Mixed costs: those for which the damage and management costs cannot be disentangled. N = number of cost estimates.

What is a non-native species? Non-native species are those moved around the globe beyond their native range, either intentionally, or as an unintended consequence of movement of people or products. As a rule of thumb, 10 per cent of introduced species become established, and of those 10 per cent become pests: low odds but sufficient to cause havoc within native ecosystems, and ranking second only to habitat destruction as an extinction risk. Non-native species have been introduced worldwide (seeds of meadow grass are amongst those transported to Antarctica in the trouser turn-ups of visitors) and some, like the special commensal case of the house mouse, are distributed globally. Commensals as invaders pose a mind-bender for the logically anguished: these animals have evolved to be fellow-travellers with people, so one can hardly call their invasions unnatural, thereby contorting the justification of killing invaders because they have been moved, by people, beyond their natural range: a house mouse's natural range is with people. A subset of commensals is domestics which, when feral, present similarly tricky logical (and ethical) strains on consistency: house cats have, de facto, domesticated humans to carry them to rich new pastures. Whether or not these species strain definitions of naturalness, they certainly become pests to biodiversity conservation (Figure 12).

When are non-natives part of the natural community? Britain has over 40 non-native bird species. (Londoners live amongst 50,000 ring-necked parakeets, apocryphally introduced by Jimi Hendrix but actually first seen in Dulwich in 1893 although, interestingly, not booming until a century later.) And it has 14 non-native mammalian species, of which five, rabbit, fallow deer, ship rat, house mouse, and brown hare, arrived about 1,000 years ago, while grey squirrel, American mink, sika and some roe deer, muntjac, Chinese water deer, edible dormouse, ferret, and brown rat arrived 100 or so years ago. Other recent arrivals—prairie dog,

12. Parakeets peek out of holes in a tree in Richmond Park, near London.

short-clawed otter, and red-necked wallaby—breed, but not much, whereas occasional escapee raccoons, raccoon dogs, skunks, coatimundis, chipmunks, and sugar gliders may survive but don't breed. For how long must a species be resident before it is accepted as part of the natural community? Fallow deer, brought to Britain by the Normans, are nowadays commonly treated as native, but muntjac deer, lagging nearly 1,000 years behind, are not. Inconsistency and prejudice are rife in media coverage: the hatred focused on American mink eating Hebridean seabirds is remorseless, while the protest at killing hedgehogs, also introduced to the Hebrides where they do something similar, is passionate. Perhaps Oscar Wilde had a point when he dubbed consistency as the last refuge of the unimaginative.

A useful idea is 'ecological citizenship'. One criterion for considering a naturalized non-native species as an ecological citizen could be the nature of the damage it causes to native biodiversity. For as long as that damage is judged intolerable then the intruder remains a pest to biodiversity, and efforts to remove it, or otherwise mitigate its impact, might be justified indefinitely. However, there comes a point where a non-native has been exerting its influence on native biodiversity for so long that a new community has emerged, to which the intruder is integral and fulfils a functional role in the ecosystem. Its removal would no longer rescue or restore the original natural state. Thus, whilst it is rational to continue acting against mink in many parts of the UK and grey squirrels (at least in Scotland) because they continue to damage native biodiversity in ways that could still be halted and reversed, in contrast it would no longer seem sensible to kill rabbits on the grounds that they are non-native, although it is understandable to do so where and when they are economic pests. The flip side is to ask, for how long must a species be absent before it loses its native status? Beavers, for example, were last recorded in the 'Cronikils of Scotland' by Hector Boece in 1526, and were rare in England by the 10th century—so they lost their foothold in the UK as long ago as brown rats gained theirs. Lynx, brown bears, and wolves were largely gone from Britain by 900, 1200, and 1700, respectively; so if their re-entry tickets remain valid, does that mean that despite their 900-year residency, fallow deer remain non-native? For reintroductions, is there still habitat for the ecosystem role they previously played to be resuscitated? For brown bears that boat has been missed, whereas for lynx it is ready to dock.

Feral domestics—a special case. Lamented introductions include the transportation, by ancient mariners, of pigs and goats to remote Pacific islands as living larders lest they were shipwrecked. Today, feral domestics threaten native biodiversity, by predation (feral goats, cats, and dogs of the Galápagos); spreading infections (avian malaria introduced to Hawaiian crows, rabies caused by feral dogs); or breeding with wild relatives (ferrets and domestic cats interbreed with polecats and wildcats). Most conservationists would favour a fiercely stringent policy of removing feral domestics—generally by killing, sometimes by birth control. The difficulty of defining Scottish wildcats genetically shows how a quest for natural purity can become highly technical and, in practice, a bit rarefied—Single Nucleotide Polymorphisms (SNPs, pronounced 'snips') sadly reveal that few if any 'pure' wildcats survive in Scotland; happily, wild-living cats that look and behave just like wildcats may have enough wild-type genes to seed a recovery.

Ecological impacts of invasive species

It is difficult to predict which invaders will have the greatest ecological impacts: outcomes are context dependent, and empirical comparisons rare—for example, fewer than 14 per cent of impact studies of invasive plants are experimental. There are five broad mechanisms of impact:

Predation. Globally, predation causes about a third of the negative impacts of invasive mammals. In Britain the invasive American mink has devastated the native water vole population of lowland rivers (a tumble towards extinction as dramatic as that of rhinos in Africa). These water voles evolved to escape otters underwater by swimming to the refuge of burrows, and to escape weasels, stoats, and even polecats that chased them into their burrows by diving out of underwater exits. But they have no effective defence against the ubiquitous predatory adaptations of American mink, which can both fit into their burrows and swim. Worse, the voles' vulnerability increased as intensification of agriculture and draining boggy ground reduced their swathes of riverside habitat to narrow ribbons, easily patrolled by mink. Britain now faces a choice between two futures for water vole conservation: American mink could probably be fully eradicated from the countryside, safeguarding water voles in perpetuity; or mink control continues forever, always fearful that shortage of funds or dedication will allow the mink population to bounce back. The first option would be vastly expensive, and raise questions about priorities for conservation funding. The second would return us to the philosophical question of the number of generations after which the American mink might become valued as an element of Britain's agricultural ecosystem.

Ecologists consider herbivory as predation on plants, and invasive herbivores are not without their impacts on other animals. They can affect vegetation (consider the feral goats eating the Galápagos giant tortoise out of house and home), and soil stability, as in the case of rabbit grazing which precipitated a landslide that threatened Macquarie Island's largest king penguin colony. In the USA feral swine, descended from farm escapees and wild boars brought in the 1900s to the USA for sport hunting, now number at least 6 million spread throughout 35 states. Their rooting and rapacious eating destroy crops, erode soil, and uproot tree seedlings, leading to deforestation; the US Department of Agriculture conservatively estimates invasive swine cause $1.5 billion of damage to crops annually.

Competition. Invasive species provide unwelcome but revealing experiments that elucidate the effects of both indirect 'exploitation' and direct 'interference' competition. In Britain grey squirrels steal tree seeds cached by reds in spring. American mink (including thousands deliberately released by Russian fur farmers to create a valuable crop) use their greater bulk to attack and drive out flimsier native European mink, which are thereby at risk of becoming the next European mammalian extinction. This is an invasive expression of intra-guild hostility. The red-eared slider is an American freshwater turtle: between 1989 and 1997, 52 million were exported worldwide as pets. Many, having outgrown their homes, were released, and now compete with native species.

Disease. Grey squirrels not only compete with reds but also infect them with the squirrel pox virus, which causes higher mortality in reds than greys. Of invertebrates, the signal crayfish introduced to Britain from North America in 1976 had, a dozen years later, colonized 250 British waterways and is now ubiquitous thanks to illegal transportation. It lives at huge population densities that erode riverbeds in a seething crustacean carpet. Worse, they carry crayfish plague, *Aphanomyces astaci*, a fungus-like infection, to which they are highly resistant but which is lethal to all of the native European species, including the British white-clawed crayfish. Feral domestics, and commensals, are notorious sources of infectious (zoonotic) disease: feral dogs spread distemper to the Serengeti's lions, and rabies to Ethiopian wolves. American bullfrogs, released from aquaria, and already invasive around the world, were found in the wild in East Sussex in 1996. The impact of these invaders on native amphibian communities is unknown, but they carry chytrid fungi that threaten amphibians worldwide. There is no room for complacency when one female can lay 30,000 eggs and disperse over several kilometres. Dutch Elm Disease, caused by the fungus *Ophiostoma novo-ulmi* and transported by Elm bark beetles, was introduced to Britain from Canada on timber in the 1960s; it killed tens of millions of elm trees.

Hybridization. When natural geographical barriers are removed, native species can be threatened by hybridization with closely related invasives. The American ruddy duck escaped collections in Britain in 1965; spreading throughout Europe it hybridizes with its endangered, native relative, the white-faced duck. The case of the Italian crested newt is similar, whereas another, as catastrophic as it is philosophically challenging, is the threat that interbreeding with domestic cats poses to subspecies of wildcat in Scotland, continental Europe, and Africa.

Ecosystem. Predation on seabirds by introduced rats on New Zealand's islands reduces forest soil fertility by disrupting sea-to-land nutrient transport in guano, affecting plants and a cascade of ecosystem processes. The 'butterfly effect' of an alien reverberating through a natural community is illustrated painfully by the near annihilation of Galápagos rice rats by invading ship rats. One species survives on a tiny corner of Santiago island, where it can rebound during El Niño years by feeding on *Opuntia* cactus that is inedible to the invaders. Moderation of the ecological impact of an invader by a third factor, such as weather, is common: the impact on British native ladybird beetles by the invader *Harmonia axyridis* is complicated by changes in mean temperature and rainfall. More generally, in the face of a warming world non-native plant species, and particularly those with an invasive potential, may out-compete natives through having earlier access to resources and pollinators.

Communities and principles

A relevant ecological phenomenon is meso-predator release. This occurs when people wipe out larger native predators, thereby also removing the suppressive influence that the larger species previously exerted to regulate the numbers of their smaller competitors. Too often, these proliferating, second-tier members of the predatory guild released from suppression are invasive species. Feral cats, for example, prosper when they are no longer subjugated by the coyotes killed off by people on Californian scrubland, or by Tasmanian devils decimated by infectious facial tumours. In both cases there was a knock-on conservation loss, of scrubland birds in Californian sage-scrub and endangered small marsupials in Tasmania.

A similar imbalance, also relevant to invasive species, is competitive release. This can occur if the control of one competitor advantages its rivals. The uneasy stalemate brokered by El Niños between native Santiago rice rats and invasive ship rats on the Galápagos archipelago was complicated when attempts to help the rice rats by removing the invasive ship rats led to an increase in house mouse numbers, with the unintended consequence that the mice then competed with the blighted rice rat.

A related ecological phenomenon is hyper-predation. This can occur when invasion by a new prey species provides food that enables an increase in native predators, which in turn impacts native prey by over-exploiting them. Thus native Californian garter snakes, fuelled by a diet of abundant introduced fish, proliferated so much that they over-hunted a native amphibian, the cascades frog.

Invasive feral pigs nearly caused the extinction of island grey foxes and threw fox and eagle conservationists into dispute. The foxes, ironically perhaps originally introduced in the canoes of native people 6,000 years earlier, are now a treasured rarity, and occur as four subspecies on only six Californian Channel Islands. In the 1990s the pesticide DDT killed the native fish-eating bald eagles, allowing mainland golden eagles to move in, feasting on the pigs but killing sufficient foxes to drive them to the brink of extinction. Conservationists faced the agonizing prospect of killing eagles to conserve foxes. The foxes were saved by a recovery programme involving killing pigs, translocating resident golden eagles, re-establishing bald eagles, and reintroducing foxes bred in captivity.

What determines an invader's success? The greater the numbers of introduction episodes, and of individuals introduced, the greater the likelihood of that species establishing. Assuming the right match of climate and resources, an invader must be able to increase from initial rarity. Track record is a good predictor: if a species is a pest elsewhere, then it is more likely to be a pest in a new setting. Incomers are more likely to gain a foothold in environments where the availability of resources fluctuates—that is, places where instability in the environmental roller-coaster periodically opens doors to opportunity. Previously islands were thought to be particularly prone to invasion, but detailed analysis failed to find any relationship between land mass size and the success of mammalian, or avian, establishment.

Blurred distinctions. In most circumstances it is a mantra of conservation that connectivity is good, because it facilitates animal movements, but that is exactly what is not wanted with respect to invasives. This is why in the pristine wetlands of Belarus, I found that water voles survived best in small, isolated ponds—exactly the opposite of the large, connected refuges that are classically thought best for population survival. Why? Probably because these isolated ponds were visited less frequently by invasive American mink.

And when it comes to connecting habitats that were formerly separate, climate change can facilitate the extension of species ranges. Cattle egrets and small red-eyed damsel flies are beautiful, and British bird- and Odonata-watchers thrill to see them. But should we regard them as undesirable non-natives brought here by carbon pollution? Do they have negative impacts on UK biodiversity? If so, can or should anything be done about it? More generally, are all invasions bad?

Prevention and prediction

A combination of globalization and irresponsibility ensures that the threat of introductions will increase. One-sixth of global land surface is highly vulnerable to invasions, including large areas of developing economies and biodiversity hotspots (Figure 13). This vulnerability is worsened by expansion of agriculture, biome shifts due to climate change, and increasing wildfire.

The drivers of biological invasions have been divided by one group of experts into six categories: (1) global abiotic environmental change (e.g. climate change); (2) global biotic change (e.g. loss and degradation of biodiversity); (3) socio-economic factors (e.g. trade and transport); (4) societal lifestyles and values (e.g. tourism); (5) scientific interventions (e.g. better prediction or mitigation); and (6) social responses (e.g. legislation). They concluded that a 20–30 per cent increase in invasives is likely in the next decades and will cause major impacts on biodiversity, driven primarily by transport, climate, and socio-economic changes.

Consider trade. The planned Chinese Belt and Road Initiative will cause a surge of transcontinental conduits, and more broadly, all transboundary projects should consider, and mitigate, the risks of invasives. Katelyn Faulkner and her team asked how the biosecurity of a country influences that of its neighbours. Their analysis is sobering. They used computer simulation to identify which of six scenarios was likely to follow if any one of 86 damaging invasive species landed in a country, considering their biology, habitats, and the economics and governance of each country in the world. This sophisticated simulation can be thought of as a giant board game, in which dice—different invasive species—land on a map of the world, and the rules dictate first whether the species would probably invade and have a negative impact, while subsequent throws would determine whether it would then spread to neighbouring countries.

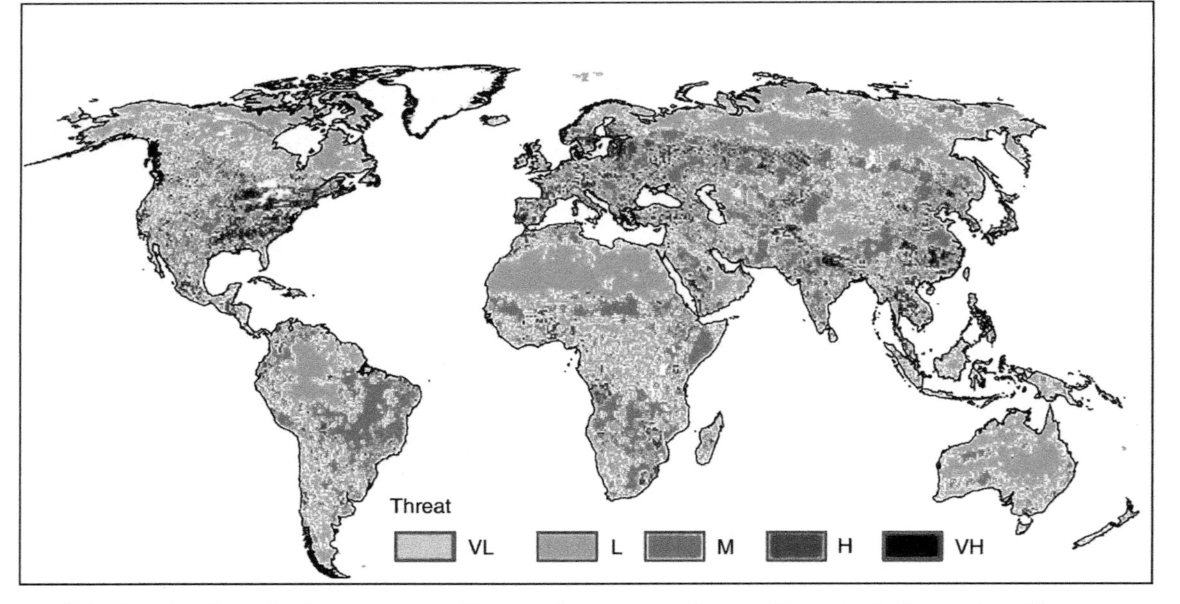

13. Global invasion threat for the 21st century. Airport and seaport capacity, as well as animal, plant, and total imports between 2000 and 2009, are combined into global introduction risk; the shading scheme runs from very high (VH; darkest) to very low (VL; lightest).

For example, a vine-like shrub, Hiptage, is native to southern Asia but the simulations suggest it could rip through West Africa. The simulations identified 2,523 feasible invasions, most of which 'would have significant negative impacts, and are unlikely to be prevented'. A third of these plausible invasions were secondary spreads onwards from a first country that had neither the capacity nor the incentive to keep the invader out (see Figure 14). The point about this model of the metastasizing spread of invasives is that preventative efforts must be implemented far from where the eventual consequences may most painfully be felt.

When is control ethical? So, when is it justified to control—which generally means to kill—non-native mammals? As a first step, I suggest that the answers to the following three questions should be 'yes' in order for lethal control to be under consideration: Does the stakeholder suffer significant loss to the 'pest' species? Does the action taken significantly reduce that loss? And is the action cost-effective?

Conservation interventions should be based on scientific evidence, but policy extends beyond science into judgement, which hopefully involves wisdom. Decisions about killing invasives, particularly vertebrates, should at least be informed (although not necessarily dictated) by animal welfare science, but judgements on the amount, and importance, of suffering are particularly difficult to make, and are rich breeding grounds for hypocrisy, or at least inconsistency. Of two small rodent 'pests' in Britain, the edible dormouse is a lot more cuddly, but a lot (2,000 years) less native than the ship rat.

Whether influenced by science or sentiment, the consequences of public perception of non-natives are far-reaching. For instance, in Italy the opportunity to prevent the early spread of invasive grey squirrels was thwarted by legal action brought by animal protesters concerned about young squirrels left to starve when

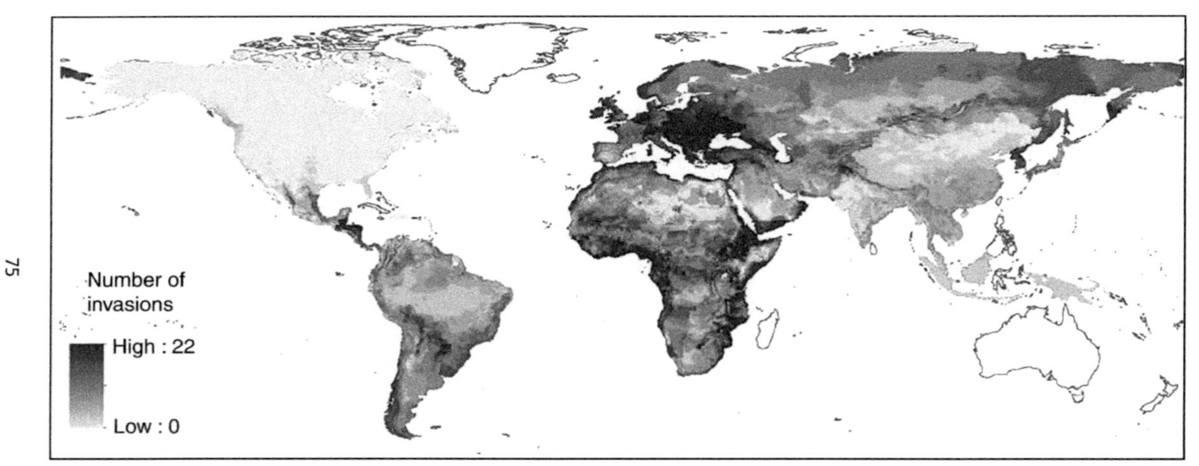

14. Regions with contiguous countries where an invasive species spreads from the country of first establishment, where it has no impact, into countries of subsequent invasion, where it has an impact.

their mothers are killed; consequently, grey squirrels have now spread so widely in Italy that removing them verges on impossible, and even reducing their numbers would involve killing many more of them—a 2022 study, delivering an immuno-contraceptive in hazelnut paste, may offer a solution.

Chapter 5
The trade in wildlife

People have bought and sold, or 'commodified', nature from fish to flora for millennia in many and varied ways. Money, or barter, changes hands when the poacher sells the rhino horn or the pet owner buys the live parrot, the lepidopterist buys the dead butterfly or horticulturist buys the orchid, when the lion farmer sells lion bones or even the opportunity to kill the lion, or the bear farmer sells bile or the hunter sells the pangolin, whether to his impoverished neighbour to make a stew or to a traditional medicine network to make a fortune. So it was, in an essay entitled *Trading Animal Lives*, that a team from Oxford's WildCRU summarized the ubiquity of wildlife trade. They noted that it may be legal or illegal, small or large scale, national or international, consumptive or non-consumptive; it may have implications, good, bad, or indifferent, for animal conservation and animal welfare, and ultimately it may be sustainable or unsustainable—all of which aspects may be fluid through time and variable in space.

Illegal trade in wild animals spans diverse supply chains, from the hunters, some impoverished, hacking off an endangered rhino's horn, abducting orphaned orangutan infants, or roasting pangolins, to mafia-like international crime syndicates that distribute and market these, often dishonourable, assets (Figure 15). Legal trade, too, encompasses the bushmeat hunter selling a porcupine at market, the captors or breeders of the millions of parrots or

15. The unacceptable face of bushmeat.

pythons traded as pets annually, or the 800 lions previously exported legally from South Africa as traditional medicinal ingredients. Even trade in domesticated animals or their products may have connections to illegal wildlife trade. For instance, the Chinese health tonic E-jiao is derived from donkey skins too often wrapped around the smuggled body parts of illegal wildlife, meanwhile depriving impoverished rural communities of the donkeys still vital for transport.

Pedro Cardoso and his team have described how the wildlife trade 'permeates the tree of life' (Figure 16). They conclude that the trade 'threatens targeted and non-targeted species, promotes the spread of invasive species, the loss of ecosystem services, the spread of diseases across geographic areas and taxa, and disrupts local to global economies especially when connected to major organized crime networks'.

Calculating the budgets of commodified nature depends on what is included. In 1997 Robert Costanza famously reckoned

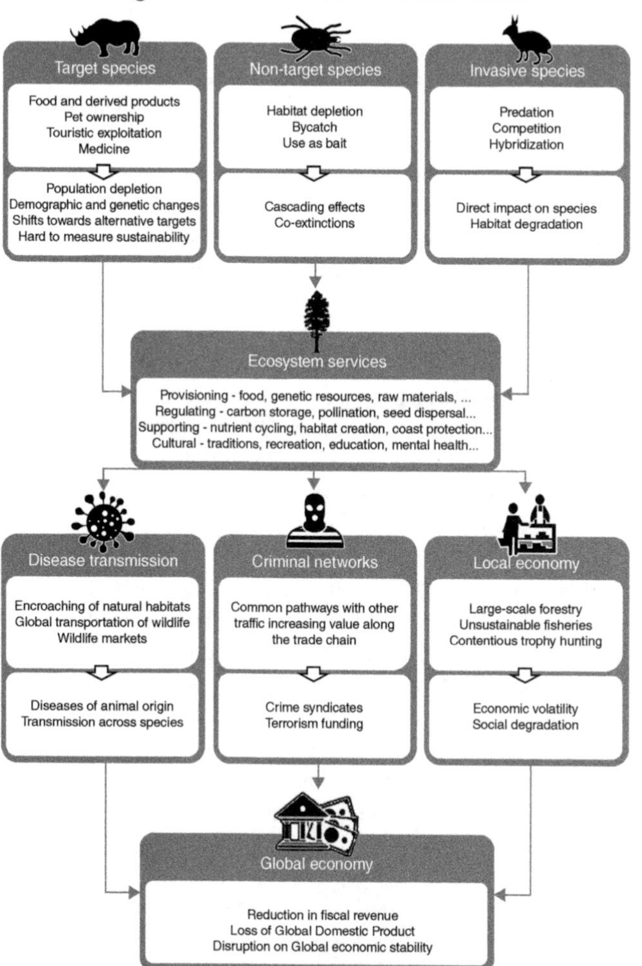

16. Causes and effects of illegal or unsustainable wildlife trade on species, ecosystems, and society.

ecosystem services saved humanity US$33 trillion annually. Even commodities less abstract than carbon, nitrogen, and water cycles clearly weigh in the profit-and-loss account: consider meat, pets, and tourism. In 2021 Dilys Roe and Tien Lee, strident voices for the sustainable use of wildlife, observed that 'trade in wild meat is worth millions of dollars to the informal economy of many developing countries across Africa, Asia and Latin America, and is also an important economic activity in many developed countries'. In 2018 it was estimated that production of 2 million tonnes of wild meat was being documented globally (doubtless the tip of the undocumented iceberg). Wild meat, excluding fish, provides protein for 150 million rural households across the global south, wildlife attractions are viewed by upwards of 6 million paying customers annually, and 47 million pheasants are, in a manner of speaking, liberated annually into the UK's shooting market. The collected monetary value of these biodiversity-based markets has never been calculated, but is most likely so large as to defy comprehension.

Limiting the undesirable aspects of the wildlife trade boils down to reducing either unsustainable supply or demand. Supply reduction may involve punitive constraints on people whose transgressions are prompted more by poverty than villainy—consider, at an extreme, ethically divisive shoot-to-kill policies towards poachers. And however thoughtful the law or strict its enforcement, reducing demand will require behaviour change on a vast scale in the face of approximately infinite demand.

Contrast two worthy aspirations, both intended to conserve wildlife: reducing those aspects of traded nature that are damaging (or seem morally inappropriate) for conservation and animal welfare; and increasing other aspects of traded nature, because in some currency (including money) this gives nature value on which its survival may depend. Which, if any, wild species it is appropriate to trade, is bitterly controversial: while advocates claim that farming rhino for their horns is the only route to saving them, opponents argue that legalizing the rhino horn trade will doom the remaining wild rhinos. Whatever the answer, the overall balance of global trade in wildlife remains more inimical than beneficial to biodiversity, its conservation, and welfare.

That an aspect of wildlife trade is legal does not guarantee that it is sustainable, nor sufficiently safeguards an animal's welfare (or human health). Between 2012 and 2016, CITES (Convention on International Trade in Endangered Species of Wild Fauna and Flora) recorded export from 189 countries of over 2 million wild animals annually, representing 1,316 species—this is only a fraction of the legal trade in live wild animals, and similar numbers are legally traded dead (estimated for 2014 as 4.5 million 'whole vertebrate equivalents'). Individual market tallies are mind-blowing—25 visits to a Bangkok market counted over 70,000 birds of 276 species—and scale up to trade involving a quarter of terrestrial vertebrate species. A 2021 analysis of 506 species of traded mammal, bird, and reptile (so ignoring fish, invertebrates, and plants from timber to orchids) estimates that 61 per cent are declining. For live wild animals traded under CITES, China was the largest exporter of mammals (98,979 animals representing 59 per cent of legal global mammal trade); Nicaragua the largest for amphibians (122,592 animals, 54 per cent); South Africa for birds (889,607 animals, 39 per cent); and Peru for reptiles (1,675,490 animals, 19 per cent). Most were en route to the USA and EU. The consequences for conservation and welfare are unknown: monitoring of source populations or provenance (wild or captive bred) is inadequate. National and international trade risks spreading invasives and zoonoses: contemplate the billions of contacts between domestic animals, people, and the tens of millions of wild animals shipped hither and thither in Asia annually. Adaptation of supply chain management and traceability tools may prove useful to reduce zoonotic disease risks from wild animal trade chains.

Legislative loopholes

Even the most basic question—what is legal and what is illegal?—can be tricky. The Indian star tortoise *Geochelone elegans* is a popular exotic pet in Thailand, but it is illegal to possess or trade them in their native India, Sri Lanka, and Pakistan. Between 2008 and 2013 Thai authorities seized almost 6,000 individuals but couldn't be sure whether they were legally captive bred or illegally traded. Similarly, wild-caught birds can be laundered into global trade from non-existent captive-breeding facilities. CITES is hampered by an incomplete roster of species, inadequate knowledge of listed species or captive-breeding facilities, and insufficient compliance, and records that are too often inaccurate, incomplete, and inconsistent.

Another regulatory complexity arises from taxonomic uncertainties: getting a species' name wrong makes a nonsense of the law. Traders in Burmese pythons (*P. bivittatus*) in China could slither around national legislation because only the Indian python was named on the List of Fauna under Special State Protection. They could also evade prosecution under CITES because CITES (correctly) lists *P. bivittatus* as native to China, so specimens confiscated in China are only in violation of CITES if they can be proven to have been trafficked across China's borders. Misnaming by skulduggerous traders can be used to similar effect, using code words or deceptive labelling. The dried paired penises (hemipenes) of protected *Varanus* lizards—for which trade is illegal—are sold by online retailers as '*Hatha jodi*', the roots of a Himalayan plant purported to have curative powers to which the lizard's sexual organs show a remarkable resemblance.

Confiscations beyond capacity

Intercepting contraband is surely a good thing, but what is the fate of the confiscates? In Yunnan, China, over 12,000 native live reptiles were seized between 2010 and 2015, of which fewer than 30 per cent were returned to the wild; the rest went to 'sanctuaries' or to pet-food. Scale that problem globally. From 2010 to 2014, confiscations reported to CITES (a fraction of all seizures) comprised 64,143 individual live animals, of 359 species, many of conservation concern. While authorities are expected to deal with confiscated animals humanely, maximizing their conservation value without promoting further illegal trade, imagine the welfare circumstances of many of these 'rescued' individuals.

Unintended consequences

Celebration of the 1987 ban on international tiger trade and, in 2007, prohibition of commercial captive breeding for their parts or derivatives, was marred for some because over 5,000 tigers were still held in 200 Chinese facilities and moreover the market turned to a problematic substitute: lions. There was a 10-fold increase between 2008 and 2011 of CITES permits enabling South African breeders to export the skeletons of captive-bred lions to Asia, linked to ethically problematic 'canned' hunting of captive-bred lions. The lion skeleton trade became a divisive debate in conservation circles, encompassing issues of entrepreneurial freedom, wildlife conservation, and the fair treatment of animals, with potential ramifications for wild lions. Now Asian demand for big cats is seeping westward, affecting jaguars killed in rural conflict in Bolivia. The general lesson of the lions-into-tigers episode is that globalization increases opportunities for product substitution, with ramifying concerns for the conservation of the analogues, and the welfare of captive-bred animals. Insofar as captive breeding might solve some problems of unsustainable trade, some species, such as pangolins, are difficult to breed in captivity and for others, conditions have been appalling: consider for instance the small wire cages housing palm civets that excrete coffee beans for tourists. A legal captive-bred supply can complicate the policing of illegal supply lines and facilitate laundering, prompting debate that parallels that concerning legalizing trade in ivory.

Substitutability is the ease with which one thing can be swapped for another. In 2008 71 per cent of consumers of tiger-based products preferred ingredients from wild tigers, and in 2011 Chinese consumers of bear bile indicated a greater willingness to pay more for wild-sourced than farmed ingredients. Of course, attitudes can change. Five criteria have been suggested to identify when wildlife farming can benefit conservation (although not necessarily welfare): when it can provide a true substitute for the wild product; meet demand and not prompt a significant increase in unsustainable demand for the wild product; be more cost efficient than the hunting/harvest of wild individuals; not rely on wild animals for restocking; and demonstrably disallow laundering.

Redirecting demand

A key hope has been to reduce demand for unsustainable wild products through education and public awareness. Sophisticated research guides the marketing of many consumer items, but initiatives to reduce demand for wildlife products have lacked understanding of customer motivations. One study revealed that would-be owners of wildlife pets were not concerned about welfare or conservation, but were put off by the risks of legal action or, most of all, catching a disease themselves. In another survey consumers of traditional Chinese medicine expressed willingness to swap to alternatives made from sustainable herbal sources. The road to reducing unsustainable trade in animal-origin medicines might lie in *redirecting* demand on to potentially less damaging products, rather than *reducing* demand through information campaigns.

Directing problematic consumption on to wildlife usages that are less damaging and more sustainable offers the hope of a win-win outcome: consumer desire is satisfied and wildlife pays for its own protection. Wildlife tourism is the leading foreign exchange earner in several countries and a multi-billion proportion of the global trillion-dollar tourism industry. There are thought-provoking comparisons between two broad—and in some respects operationally similar—categories of wildlife tourism, based on photography and hunting. Both are huge global industries. Wildlife tourism generates US$120 billion globally annually. Rwanda's mountain gorillas generate US$17 million annually in permit fees alone; South Africa's Isimangaliso Park provides 7,000 permanent and 3,000 temporary nature-based jobs; in 2016, visits to the USA's park system generated US$34m and 318,000 jobs; and jaguar tourism to the Pantanal generates US$6.8m p.a. (set against US$121,500 in stock losses).

One variant of photo-tourism, illustrating the potential for good and the risk of bad, is Wildlife Tourist Attractions (WTAs) that offer opportunities to interact with non-domestic animals, in captivity or in the wild. On the down-side, globally, 24 types of WTAs, ranging from riding elephants to cuddling koalas, estimated by conservationist Tom Moorhouse to host 3.6 million tourists annually, negatively impact the welfare of between a quarter and half a million individual animals, and are inimical to the conservation status of the species in the case of 120,000–340,000 animals. Many WTAs involve animals captured from the wild, and thereby lost to conservation as surely as if they had been eaten or hunted. On the up-side, six WTA types, affecting 1,500–13,000 animals, had net positive impacts on both species' conservation and individual animals' welfare.

Hunting for table or trophy is amongst the largest commodifications of biodiversity; there are 1 million duck shooters and 11 million deer hunters in the USA, and 7 million hunters in Europe. It raises unexpected, sometimes uncomfortable, parallels, as between pheasant shooting in the UK and lion shooting in Africa—both include put-and-take variants, and both can be argued to deliver conservation costs and co-benefits. These two forms of sport hunting share features likely to be influential for the future of sport hunting more widely, including the extent to which sport hunting maintains land for wildlife (game bird management has motivated the creation, retention, and maintenance of many lowland woodlands); the impacts of intensification (e.g. the extent to which quarry are reared and released); and concern for the welfare of quarry animals. Having been, in my early career, drawn into the wrath of Britain's fox hunting debate, I was plummeted into the similar ethical impasses of lion hunting when a lion we were GPS-tracking, nicknamed Cecil, was bow-hunted. Over 4 million people visited our website the night the story broke, and in the following weeks controversy raged in over 12,000 print-media articles in 125 languages. This was arguably the most-reported conservation-relevant story in history and prompted me to ask: can the Cecil *moment* become the Cecil *movement?* And to wonder whether future generations might regard well-intentioned utilitarian conservation arguments for trophy hunting as similar to those once invoked for child labour. As we concluded in our analysis,

> it may be that the preoccupying interest in Cecil displayed by the millions of people who followed the story betrays a personal, and thus potentially political, value not just for Cecil, and not just for lions, but for wildlife, conservation and the environment. If so, then for those concerned with how wildlife is to live alongside the human enterprise, this is a moment not to be squandered and one which might have the potential to herald a significant shift in society's interaction with nature.

Certainly, the reverberations of Cecil's death (which was itself not unusual) were seismic; 40 airlines stopped transporting hunting trophies. The incident prompted debate, fuelled by knowledge gaps such as the causes of lion mortality, the amount of land used for lion trophy hunting, the extent to which trophy hunting depends on lions for financial viability, and the vulnerability of areas used for hunting to conversion to land not used for wildlife if trophy hunting ceased; and also by ethical dilemmas, leading to the first formal analysis of the pros and cons of trophy hunting. The latter explored various premises—for example, because lions possess intrinsic value they should not be killed without good reason—which exposed a topic of scholarly complexity that, revealingly, was almost completely ignored in the babble of subsequent tribally polarized debate. In short, a position associated with utilitarianism acknowledges that properly regulated trophy hunting, like it or not, is a part of current African wildlife conservation, with the biodiversity gain outweighing the loss of individuals as trophies. Big cat hunting is only a fraction of predator hunting, and, by number and finance, a tiny, although to me particularly distasteful, part of hunting globally. When asked in a report to the UK government whether I would favour a ban on lion hunting (a highly simplistic question), mindful of unintended consequences of impetuosity, I replied that I favoured a journey rather than a jump. The fear is that those who would instantly ban trophy hunting might, in jumping to the moral high ground, find there may be rather less African wildlife to be seen from that lofty position.

Enforcement

Markets for wildlife trade are susceptible to regulation insofar as it's easier to exert control on hubs. Consider the ban imposed by the Standing Committee of the National People's Congress on 24 February 2020, endorsed by President Xi Jinping. This resulted in the almost overnight closure of 12,000 wildlife-related businesses and deletion of almost a million online sources of information linked to wildlife trade (and, importantly, the disbursement of about US$1 billion to compensate the wildlife farmers whose businesses had evaporated). This was a remarkable silver-lining to the COVID pandemic. And Chinese authorities cracked down on nearly 12,000 wildlife crime cases in three months during a multi-departmental '2022 Qingfeng Action' law enforcement operation, which targeted illegal wildlife trade at various stages, including harvest, purchase, transport, import/export, and resulted in more than 100 million yuan (~US$14.8 million) in fines. Where there's a will, there's a way.

On the other hand—and the realities of conservation always involve another hand—in 2021, wildlife farming in China was worth US$8 billion annually, and it contributes to poverty alleviation. This convinces some advocates of sustainable use to favour improved biosecurity over a ban on wild meat markets. They fear a cascade of unintended consequences if consumers of wild meat turned to domestic meat, itself a burgeoning source of infectious diseases, with resulting increases in habitat loss, and thus species extinction, in the quest to feed domestic stock. Before you know it, pondering the merits of regulating cane rat farms bounces you into a discussion on global vegetarianism.

Modernity means living through the internet—this, for example, is how most saiga horn is now traded. Online marketplaces and social media sites are used by both legal and illegal traders efficiently to procure wildlife and to expand their consumer base, with many sales made directly over social media. For instance, in 2016, 22–46 per cent of international social media posts trading orchids pertained to wild-collected plants. The online wildlife trade presents new challenges: closed social networks, difficult-to-trace third-party financial transactions, and increasing use of cryptocurrencies; but also new insights into people's activities, interests, and emerging markets. Online wildlife trade activities are currently neither monitored nor regulated by hosting platform companies and there are no regulatory 'watchdog' institutions; monitoring is generally by NGOs.

Extractive sustainable use is becoming unrealistic

The UN's 17 Sustainable Development Goals (SDGs) and fundamental principles, notably poverty alleviation, attention to international law and human rights, and appropriate, sustainable custody and use of nature, are timelessly worthy. But the current reality of human use of commodified natural products is of rampaging forest loss, radical reductions in the population size and genetic diversity of apex predators, roads penetrating into every wilderness, and exponentially rising numbers of human users, utilizing increasingly powerful and efficient technology, impacting an exponentially shrinking wildlife resource, subject to an increasing number of compounding pressures. These trends do not undermine the principle of sustainable use, but raise the spectre that the number of contexts (species, places, forms of exploitation) where resources are sufficiently intact and resilient to make extracting them sustainably a realistic option is shrinking.

While the rustic notion of a rabbit for the pot may be sentimentally appealing, 'bushmeat' illustrates the scale of unsustainability: from a single region of west Africa (35,000 km^2 of the Senaga-Cross forests of Nigeria and Cameroon), approximately 800 kg of wild meat was extracted per km^2 annually; that's close to a million reptiles, birds, and mammals sold. Across the tropics, an estimated 6 million tonnes of animals (mostly ungulates and rodents) are extracted every year. Extraction at this scale is unsustainable, particularly for large-bodied and slowly reproducing species.

The vision of extractive sustainable use is threatening to become a mirage. Human population sizes and rates of usage have risen so quickly, and the number of species and habitats in which consumption could be sustainable at scale has dwindled so rapidly, that the concept surely begs for re-evaluation. It is surprisingly controversial to suggest that the onus should be on traders to demonstrate that wildlife use (and e.g. CITES certification) is sustainable, rather than on conservationists to demonstrate that it is not.

What to trade?

The debate over which wild species should be traded is fierce; approaches fall into two broad categories: make trade illegal or allow some or all trade to be legal and seek to manage it sustainably. A team led by Liz Bennett, a wise veteran of this battlefield, arrived at the unsurprising conclusion that the answer should be specific to particular cases, considering species-specific attributes of biological productivity, management context, and demand. Considering examples ranging from the gathering of fungi in Mexico to wildlife ranching in South Africa, one 2021 study evaluated how wildlife trade contributes to the UN SDGs; it concluded that prohibitions applied with insufficient thought could imperil some SDGs, with little benefit (e.g. for public health, in terms of preventing a future pandemic).

In Chapter 3 I pondered how difficult it is to answer the question: what is biodiversity conservation? In the context of trade, the IUCN offers one definition:

> Conservation is the management of human use of the biosphere so that it may yield the greatest sustainable benefit to present generations while maintaining its potential to meet the needs and aspirations of future generations. Thus, conservation is positive, embracing preservation, maintenance, sustainable utilization, restoration, and enhancement of the natural environment.

Even against this notably anthropocentrist definition, biodiversity conservation is failing. Despite excellent applications of the sustainability paradigm, such as Marine Stewardship Certified fisheries and Blue Ventures, global rates of wildlife utilization are unsustainable. If sustainable wildlife trade is a correct use of wildlife, it must change—and urgently—to work for both people and wildlife. Amidst overheated, if understandable, demands to ban all wildlife trade, thoughtful authors began to express caution, bearing in mind the complexity of the links between this trade, legal and illegal, and human well-being, warning of unintended consequences for some of the world's most vulnerable people. One size of solution is unlikely to fit all. It will take political bravery and a firm grasp of the pros and cons to fortify the regulation of wildlife trade and food supply, along with the realization that the use of animals by communities around the world may need to evolve in line with societal expectations consistent with a new understanding of risk and restrictions in the post-COVID-19 world. This brings us to the interface of biodiversity conservation and the risks of infectious disease.

Chapter 6
Wildlife disease

Wildlife diseases are relevant to biodiversity conservation because they can threaten people and wildlife, including endangered species. In both cases, whatever the benefits, attempts to control the disease can perturb the behaviour of survivors, threaten populations, and alter their dynamics, prodding a hornet's nest of operational and ethical issues. Wildlife diseases can have major ecological, socio-economic, and health consequences, so they prompt intervention—often lethal—gravely affecting whether wildlife is thought of as friend or foe.

Zoonotic diseases—that is, diseases caused by a pathogen that has jumped from animal to human, and vice versa—cause millions of human deaths annually. Moreover, over 70 per cent of emerging human diseases originate from exposure to wildlife through trade. The opposite risk is exposing wildlife to disease, for example, when tourists visit the famous Rwandan gorillas. Diseases such as HIV, Ebola, severe acute respiratory syndrome (SARS), avian influenza, and, most recently, COVID-19 cause economic losses globally, and strain international relations. Even before COVID-19, global economic damage caused by emerging zoonoses totalled hundreds of billions of US dollars over the two decades 2000 to 2020. In 2020 the IMF estimated total global economic damage from COVID-19 through to 2025 would amount to trillions of dollars (in the UK the expected cost is £368 billion, or £5,500 per person).

Fears of disease threats are sometimes misplaced. The crab-eating fox (which is neither a proper fox nor does it much eat crabs) was a suspected wildlife (sylvatic) reservoir of *Leishmania chagasi*, which causes visceral leishmaniasis, a major public health problem in Brazil. Orin Courtenay and I radio-tracked these foxes in Amazonia, and found they ventured close to villages. Working with epidemiologist Chris Dye, we found that the foxes were infected with *Leishmania*. But in 2002 Orin, Chris, and others reported on the immunology and serology of 26 wild-caught foxes exposed over 15 months to laboratory colonies of the sandfly vector *Lutzomyia longipalpis*: 78 per cent of these foxes caught the parasite, although none got ill. Crucially, following 44 sessions when sandflies banqueted on foxes, not a single engorged sandfly, of 1,469 dissected, had been infected by the parasite (in contrast, 11 per cent of sandflies fed on co-occurring, or 'sympatric', infected dogs became infected). The ensuing mathematics indicated that domestic dogs were responsible for at least 91 per cent of disease transmission, and the crab-eating foxes were off the hook.

Disease fundamentals

Despite the myriad of pathogens and hosts, and the variety of their interactions, the epidemiological complexity distils down to several intuitive general principles. One is embodied in a parameter, R_0, widely familiar thanks to the COVID-19 pandemic. R_0, known also as the basic case reproduction number, is the average number of new individuals infected by a single infectious individual. The logic of R_0 is simple: if one primary case of infection generates more than one secondary case on average ($R_0 > 1$), then infection will spread through a population of hosts. The challenge for infection and disease control is equally clear: one primary case should generate only one or fewer secondary cases on average: then, sooner or later, infection will disappear from the host population. Depending on the pathogen and the host, this might be achieved by isolating infectious individuals and their contacts (quarantine), curing infection by treatment, preventing infection by vaccination, or, commonly in the case of wildlife, killing the wild animal host.

If $R_0 > 1$ and high, the disease is likely to rip through the population, causing an epidemic, but may run out of susceptible hosts to infect, as can happen with rabies. If $R_0 > 1$ and low, the disease will be less severe but more persistent, as the hosts would have time to breed or emigrate, thereby maintaining the number of susceptibles available for contact. The result is endemic disease; viral hepatitis, herpes, and COVID-19 all follow this pattern. The factors that affect disease dynamics include the agent of transmission (be it a micro- or macro-parasite, bacterium, or virus) and the method of transmission. This latter may be density dependent, as in the case of distemper in lions, or contact dependent, for example, Mustelid Herpes Virus—a sexually transmitted disease of badgers.

Crucially, species have evolved immune systems, giving some greater immunity to certain pathogens than others. Some pathogens can infect only limited taxa; some parasites are host specific, while others are versatile. Selection for immunity is a fundamental evolutionary mechanism driving speciation—the splitting off of populations to form new species. Indeed parasites merit conservation because they sustain ecological processes, including immuno-genetic selection through mortality and mate choice. Hosts and parasites evolve hand in hand. Across the mammalian tree-of-life, we see a correlation between diversity of ectoparasite species and the number of species into which the mammal group has diversified. Similarly, in the case of birds, there is evidence that haematozoan blood parasites exert strong natural selection, mediating the maintenance of genetic diversity. If we take an egalitarian view of biodiversity, why should a host be more valued than its parasite? Should we be sadder about the passenger pigeon, or the mite that it apparently carried to oblivion? And is it important to reintroduce beavers to Britain in order to restore their parasites, particularly their unique demodectid mite, *Demodex castoris*, a newly discovered species?

The far-reaching role of pathogens, and specifically parasites, in natural populations came to be appreciated following the theoretical work of Roy Anderson and Robert May in 1978. It was obvious that pathogens cause illness, even death, to individuals, but they showed that this process can affect populations, regulating numbers through their effects on fecundity and mortality. At the population level, diseases can cause dramatic die-offs; in central Kazakhstan in 2015, some 200,000 saiga antelopes died from haemorrhagic septicaemia from the bacterium *Pasteurella multocida*. In other cases, diseases may have sub-lethal effects with long-term impacts, as for example in the case of the gastrointestinal roundworm *Ostertagia gruehneriand*, a parasite of Svalbard reindeer that depresses their body mass, back fat, and fecundity. Sick prey are critically susceptible to predators, and at the individual level, character matters: bold hosts with an active personality may risk greater exposure to disease than the meek. So by these means, the parasite's presence creates selection pressures. Parasites can also directly alter their host's behaviour. An eerily manipulative relationship exists between brown rats and one, but only one, of their numerous parasites: *Toxoplasma gondii*, which can infect all endothermic vertebrates but has only one definitive host—the cat. Therefore *T. gondii* inhabiting a rat need to find a cat to complete their life cycle. Astonishingly, rats infected with *T. gondii* show more cat-attracting behaviours than uninfected rats. Infected rats actively seek out the scent of cat urine: the parasite alters the behaviour of these suicidal rats, driving them into the cat's jaws. There is no difference between rats infected with *T. gondii* and those that are uninfected in other behavioural traits that don't benefit the parasite, such as social status or mating success. The rat's fatal cat-seeking behaviour is driven by the parasite in order to enhance its own transmission.

Disease can also seriously impact animal populations indirectly through deleterious effects on other, ecologically linked, species—a dramatic case is the sylvatic plague that, after introduction into San Francisco in 1900, spread north and east, wiping out entire colonies of prairie dogs, thereby starving, as well as directly infecting, black-footed ferrets. The ferrets were reduced to five breeding females (though thanks to a remarkable captive-breeding programme, including reproductive cloning using a technique called somatic cell nuclear transfer to increase genetic variability, there are now about 30 reintroduced populations).

When is disease relevant to conservation?

A purist might say that infectious disease, as a natural process, falls outside the remit of conservation, just as it is not the conservationist's role to intervene in natural predation. Clear-cut exceptions include where there is zoonotic risk to human health or where the impact of disease on biodiversity, in the form of threatening species diversity, endangerment, or ecosystem function, is affected by factors traceable to people. For example in 2022 in South-East Asia, African swine fever threatens 11 endemic species of wild pig, including bearded and warty pigs, undermining the tigers and clouded leopards that prey on them, and a plethora of ecosystem processes such as soil tillage, seed and nutrient dispersal—all this in the remotest wildernesses but due to a viral jump from Chinese domestic pigs in 2018 (Figure 17). If the survival of species, or populations, becomes threatened primarily by disease, the argument for intervention is strong. The possible interventions include isolation (e.g. Norway's use of fencing to combat chronic wasting disease among reindeer), habitat modification (draining wetlands where avian cholera occurs), and genetic manipulation (e.g. to manage the genetic resistance of frogs to chytridiomycosis). But traditionally, the first resort, not the last, has been killing, in the not always realistic hope of driving down R_0. Vaccination can offer a more sophisticated option, but it is not always practicable, as in the case of Ebola in gorillas, and it

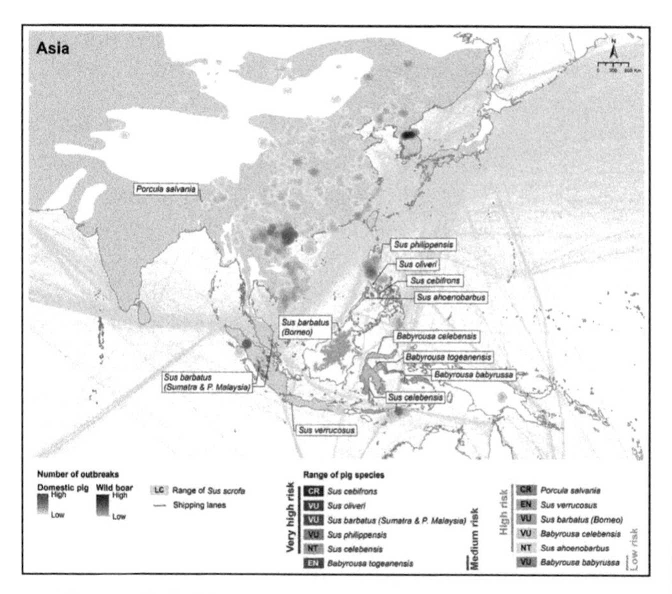

17. The spread of African Swine Fever threatens South-East Asia's 11 wild pig species. Islands of denser shading correspond to larger outbreaks (more pigs infected) as of early August 2020. The ranges of wild pig species in Asia are shown with solid shades. Letter codes within the pig species legend denote IUCN threat level, listed here in increasing order: Least Concern (LC), Near Threatened (NT), Vulnerable (VU), Endangered (EN), Critically Endangered (CR). Very high to low risk assessments justified in the original paper.

can have complications. For instance, if lions proliferate following Canine Distemper vaccination, cheetah might be disadvantaged.

Perturbation and vaccination

After a doctorate on red fox behaviour, I began my career by nervously addressing the top brass of the World Health Organization on fox-borne rabies. My pitch to these senior vets who, to parody only slightly, thought the only good fox was a dead one, was that foxes were members of societies whose disruption by killing could increase disease transmission amongst the survivors:

a *perturbation effect* where good intentions make things worse. Rabies, a viral infection generally injected in saliva via a bite, is almost always fatal to people and to red foxes (although interestingly less so to mongooses or raccoons). Until the mid-1970s the prevalent view was that rabies could be controlled by killing sufficient vectors—in Europe, the fox—to reduce R_0 amongst the survivors below 1. Gassing the dens of hundreds of thousands of foxes in the 1950s and 1960s—wiping out badgers as collateral damage—neither eradicated nor slowed the spread of European rabies. Following a 1961 trial by George Baer of the Centre for Disease Control in which cyanide was replaced with vaccine launched into their mouths by sinisterly-named 'coyote-getters', the breakthrough came from Franz Steck and others in Switzerland, where rabies arrived in 1967. Franz's trial involved Y-shaped valleys: as rabies spread up the stalk of the Y, the entrances to each arm of the fork were bombed with chicken heads, laden with a biomarker detectable in the teeth of foxes that ate them. At the mouth of one branch of the fork the baits contained poison; at the other, a live attenuated oral vaccine named SAD. The outcome proved to be happy: rabies leapt over the poisoned *cordon sanitaire* but in the other branch the disease stopped dead when it hit the undisturbed social system of the vaccinated foxes. Franz was killed in 1982 when his helicopter crashed while dropping chicken heads. His memorial was to eliminate the disease rather than its wildlife host: by 1996 Switzerland, 2.6 million baits later, was rabies free.
Sixteen countries began oral vaccination by the 1990s, and European fox rabies may soon be history as a result of combining knowledge on fox behavioural ecology and advancing vaccine technology. (Final eradication has faltered, due to people moving animals illegally, lack of funding commitment, policy in Eastern Europe, and the arrival of invasive raccoon dogs.)

Rabies may not extinguish the ever-resilient red fox, but it could mean the loss of the cliff-dwelling Blanford's fox, with powder-puff tail—one of south-west Asia's rarest predators. In the

1980s these foxes were discovered at Ein Gedi, Israel, far west of their known range, but no sooner were they discovered than they were facing extirpation by a rabies outbreak in co-occurring (sympatric) red foxes and an associated fox-cull. Rabies also threatens extinction for the Ethiopian wolf, as rare as it is beautiful. Claudio Sillero has devoted a lifetime to the conservation of the remaining 500 or so individuals as they retreat from agricultural expansion and climate change to ever tinier mountain plateaux where, in their last bastion, the Bale Mountains, they are pummelled by rabies outbreaks. These outbreaks have caused up to 77 per cent mortality every 5–10 years. Modelling suggested that target vaccination of packs at strategic localities, such as the mouths of movement corridors, would solve the problem.
It worked—although catching and injecting each wolf was a labour of love. In 2016, the team tested four baits carrying a biomarker that would be detectable later in the wolves' blood in 12 packs, using three delivery methods, the best of which proved to be a horseman approaching a targeted wolf by night, and throwing the steak to it. They also baited three packs with a modified live attenuated virus vaccine named SAG2. A fortnight later, 86 per cent of wolves that were revealed by the biomarker to have eaten vaccine-laden bait had developed adequate protection against rabies. All save one of the vaccinated wolves were thriving 14 months later. A large-scale preventative vaccination and monitoring campaign now looks feasible.

The perturbation effect has been most prominent in the context of bovine tuberculosis—an international problem affecting *c.*50 million cattle worldwide and costing $3 billion annually. In the UK, since the discovery in 1974 of a tuberculous road-killed individual, European badgers have been considered a reservoir for the disease. Bovine tuberculosis (bTB) is caused by the bacterium *Mycobacterium bovis*, which heavily impacts dairy cattle, but rarely causes morbidity or mortality in badgers. Whether badger culling intended to reduce bTB in cattle has been successful is controversial, and one explanation of its equivocal outcomes is perturbation. The Random Badger Control Trial, the biggest, most expensive wildlife management experiment ever attempted, compared 10 replicates of three treatments—killing badgers reactively where the disease broke out in cattle; annihilating them proactively; and doing nothing—each applied to 100 km^2 of farmland. The Reactive treatment was abandoned because it was making things worse, as predicted by the Perturbation Hypothesis. In the case of the Proactive treatment, while culling was occurring, the number of confirmed herd breakdowns within the culling area was 23 per cent lower than in non-culling areas. However, herd breakdowns increased by 25 per cent in the 2 km-wide perimeter surrounding the core culling area. After nine years, five years after the trial ended, the proactive treatment was associated with an overall reduction of 26 per cent (with wide variation) in cattle disease in the removal areas. That overall gain was experienced very differently by farmers in the centre of each removal zone and those at its edges, where the improvement averaged only 8 per cent with wide variations, ranging from an overall worsening of 16 per cent and an improvement of 35 per cent—a dubious gain for so much effort, expense, and killing. The Expert Group concluded that 'badger culling cannot meaningfully contribute to the future control of cattle TB in Britain', and that 'some policies under consideration are likely to make matters worse rather than better'. They also remarked that 'weaknesses in cattle testing regimes mean that cattle themselves contribute significantly to the persistence and spread of disease in all areas where TB occurs'. Nonetheless, faced by a powerful agricultural lobby, the UK government has continued to kill badgers. A comparison in 2022 of the incidence of bovine tuberculosis between culled and non-culled High Risk Areas detected no difference between them. If anything, in most years, the incidence levels in cattle were slightly higher in areas where the badgers had been culled (Figure 18).

This analysis was met by a vehement rebuttal from Defra (the UK's Department for Environment, Food & Rural Affairs),

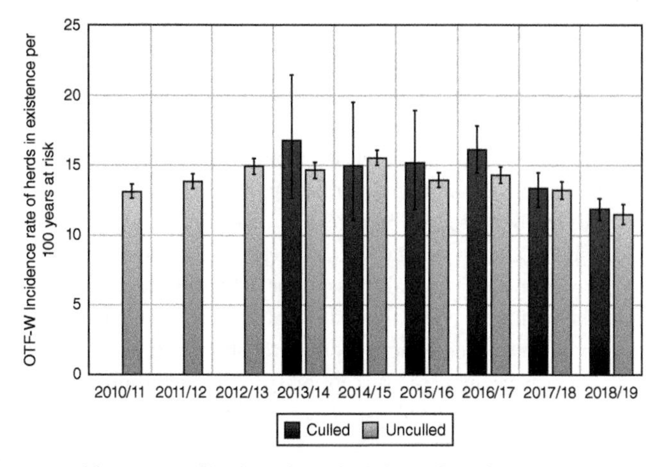

18. Incidence rate of bovine tuberculosis in cattle within and outside 30 badger cull areas of the High Risk Area of England, during badger cull years (September to August) 2013/14–2018/19.

with each side pointing to analyses that favoured their interpretation. By then an estimated 140,000 more badgers had been killed without noticeable effect and the hopes of a generation of dairy farmers cruelly misplaced. One can only wonder whether an effect that is so hard to demonstrate has been worth the cost of bringing it about. A view widely held amongst conservation biologists is that the greatest single contribution to solving the bovine tuberculosis problem lies in more effective testing for, and early culling of, infected cattle, and hopefully success in the logistically complicated process of vaccinating cattle and badgers—both of which are immunologically sophisticated. Trials are currently under way to accelerate towards planned deployment of a cattle tuberculosis vaccine by 2025.

Wider considerations

Disease also affects reintroduction of animals and plants. Recall that water voles are threatened in Britain by invasive American mink. The bacterium, *Leptospira*, that causes Weil's Disease in humans, also threatens water voles, delaying sexual maturity and reducing growth and survival rates. During reintroduction, captive-bred voles had been disease free on release, but after four months in the wild, 43 per cent were excreting *Leptospira*, compared with 6.2 per cent in natural populations. Stress in wild animals can be measured by a technique called Leukocyte Coping Capacity (LCC); lowered LCC scores suggest immuno-suppression and therefore physiological stress. Prior to their release, it had been convenient to house the young water voles in groups, but this lowered their coping capacity scores, probably due to social stress, which may have increased susceptibility to the bacterium. Henceforth, the young voles were housed separately.

Ecology, and thus conservation, is awash with cascades and unexpected connections. Tasmanian Devil Facial Tumour Disease (DFTD) is a non-viral infectious cancer transmitted by biting. First seen in 1996, within 20 years it had spread to 80 per cent of the Tasmanian devil's geographic range, causing population crashes of more than 95 per cent. This decimation of devils caused increases in the population of feral cats they had previously kept down. Worse, as devils crashed and cats soared, the smallest local meso-predator, the eastern quoll, rapidly declined. Quolls can catch the parasite *Toxoplasma gondii* from the cats. Quoll numbers may also have plummeted due to a run of wet, warm winters, and cat predation on their young may now prevent their recovery—a 'predator pit' that could worsen if cats, previously diurnal to avoid nocturnal devils, become more nocturnal and thus encounter young quolls more. Emphasizing the importance of genomics to conservation, recent evidence of selection for alleles (gene variants) resistant to DFTD amongst the wild population raised concerns that insurance population devils, isolated from the wild, may be unsuitable for future translocations. Thankfully, comparison of both genome-wide and functional diversity (in over 500 genes) of Tasmanian devils has shown this not to be the case. Ecological complexity necessitates a case-by-case approach to the impact of disease on biodiversity, and that requires monitoring, surveillance, and evaluation.

Disease intersects with climate change, which is producing shifts in ranges of animals. It has caused the northward movement of the meningeal worm of white-tailed deer, which now encounters moose and red deer at these higher latitudes, and is causing the wider spread of lungworms in muskoxen. Mosquitoes are also on the move, and with them, malaria. The intersection between disease and climate led to warnings that managers of national parks and protected areas should plan carefully to prevent the arrival of infectious disease and prepare for its management. Spreading to higher latitudes as a result of climate change also intersects with disease in amphibians: when conditions cool, amphibians from warm climates experience worse infection by chytrid fungus than do those originating from cool regions.

Quick thinking

Infectious disease may compel conservationists to act fast, on imperfect information, without knowing the size of the relevant population or the number of infected individuals. A 2021 review suggests harmonizing disease and population monitoring in wildlife, so that passive disease surveillance morphs into something more holistic, what is known as integrated wildlife monitoring (Figure 19). Such joined-up thinking aligns with the One Health approach that highlights the intersection of people, other animals (domestic and wild), and the environment.

The deep links between humans, domestic and wild animals, and environment is evident in another tragic instance of wildlife disease: the 2002 SARS outbreak infected 8,096 people, killing 744. The response in Guangdong Province, China, had been to kill thousands of masked palm civets, along with raccoon dogs, ferret badgers, and domestic cats.

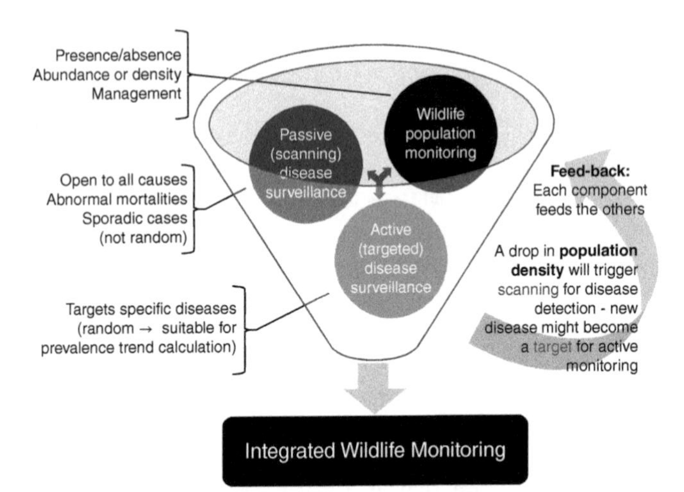

19. Integrated wildlife monitoring as the combination of population monitoring, passive (scanning), and active (targeted) disease surveillance.

On the heels of SARS, the COVID-19 outbreak is but one of many pandemics caused by viruses jumping from non-humans: HIV, the cause of AIDS, derived from the butchering of wild chimpanzees as meat; the 2009 novel H1N1 influenza virus passed from pigs to humans at a meat production facility in Mexico. The repeated plagues in medieval times were most likely spread by the *Yersinia pestis* bacterium associated with rats and their fleas; cohabitation with rats has killed hundreds of millions of people. In 2022, a highly divergent lineage of SARS-CoV-2 in white-tailed deer, possibly capable of deer-to-human transmission, was found. The message for biodiversity conservation and human health is the same: these risks are long-standing, worsening, and demand holistic attention.

The source of SARS-CoV-2, the virus behind COVID-19, probably leads back to a remote copper mine in Mojiang, Yunnan Province, and via some reshuffling of the viral RNA, to another virus named RaTG13, whose genome was a 96.2 per cent match to

SARS-CoV-2 and which infects Horseshoe bats there (the Ra stands for *Rhinolophus affinus*—the name of the bat). An early thought was that pangolins might have been intermediary hosts, but the genomics contradicted this; indeed, we had been working with Chinese colleagues in those Wuhan markets and found that neither pangolins nor bats had been traded there, although a staggering 47,381 individuals from 38 species, including 31 protected species, had been sold between May 2017 and November 2019, some farm bred, others, to judge by snare injuries, wild caught. In 2022, a team led by Mike Worobey examined grime left in a cage once occupied by raccoon dogs and found a match to the virus strain that had infected the first hotspot of human patients in the vicinity of the same market stall. Winston Churchill popularized the aphorism, 'Those that fail to learn from history, are doomed to repeat it.' Nobody, least of all conservation biologists, will be surprised that the continuation of wildlife markets in China allowed another and more infectious Coronavirus to emerge.

While the 6 million+ human death toll from COVID-19 is huge, these are not the highest zoonotic Case Fatality Rates: in West Africa in 2014–16, the Ebola virus reportedly killed 11,310 of 28,616 infected people (bats are the reservoir, but Ebola passed through monkeys and duikers en route to humans), whilst since 1981 around 32 million people have died of AIDS. While infectious diseases may have passed in either direction between humans and other animals for centuries, the pressure on the environment of human populations today, their encroachment into every wilderness, together with the wildlife trade, has increased the risks of zoonotic disease immensely to create what, in an essay in 2006, I warned was '*an explosive mix of risks*'.

Chapter 7
Human–wildlife conflict

Legend has it that in AD 750 Guru Rinpoche flew astride a tigress to deliver Tantric Buddhism to Bhutan. Later, in 1993, when 5,000 pastoralists were encircled by the designation of the Jigme Singye Wangchuk National Park they were therefore warmly disposed towards the tigers therein. However, their happiness—remember Bhutan wisely measures itself by Gross National Happiness—began to fade when the Forest and Nature Conservation Act banned killing wildlife and restricted their use of this forest. Camera-traps revealed that the new park contained eight tigers and by 2019 conservationist Ugyen Penjor wrote that Bhutan was the tiger's most viable refuge. While the world might rejoice, the 20 per cent of farmers reporting that predators killed their stock, who thus lost 84 per cent of their annual income, were less happy. Indeed, 44 per cent of tiger scats and 73 per cent of leopard scats contained livestock remains, and more than half of villagers blamed the park for their hardships. However, farmers weren't blameless: husbandry was inadequate and 60 per cent failed to construct adequate night-time corrals. This illustrates human–wildlife conflict (HWC), but conflict, rather like a famous US Supreme Court ruling on obscenity—'I know it when I see it'—is easier to recognize than to define. Nonetheless, the IUCN-HWC and Coexistence Specialist Group proposes:

> Human-wildlife conflicts are struggles that emerge when the presence or behaviour of wildlife poses actual or perceived, direct and recurring threats to human interests or needs, leading to disagreements between groups of people and negative impacts on people and/or wildlife.

So where people meet nature, especially the sort that's red in tooth and claw, and even when they are culturally well disposed to it, there is often conflict.

In 1785 Scottish poet Robert Burns, ploughing his field, destroyed a wood mouse's nest. He addressed the evicted mouse with the famous words 'wee sleekit cowrin timrous beastie', in a poem that laments, with unsurpassed empathy, that struggle. Too often HWC is framed in terms of the most charismatic megafauna living alongside the poorest people, but even a contemporary wood mouse would recognize her ancestor's conflict with agriculture (now worsened by chemical sprays). HWCs—an intensifying threat to numerous species—are described by IUCN as '…one of the most complex and urgent challenges for wildlife conservation around the world', whilst the World Wildlife Fund (WWF) observe that HWC is '…also a significant threat to local human populations [and] if solutions to conflicts are not adequate, local support for conservation also declines'. HWC is a conservation industry, with an exponential increase in 2,197 HWC publications in 320 journals from 128 countries between 2003 and 2021.

These publications focused on three facets of HWC: conflict between humans and carnivores, between humans and herbivores, and protection of the human dimension. Over the last decade the focus has shifted from the conflict itself to the coexistence of humans and wildlife.

HWC is almost infinitely variable—some people tolerate losses from wildlife despite living in poverty, others experience negligible economic losses but are relentlessly intolerant. This kaleidoscope of complication is further muddied because emotion rarely matches straightforwardly the severity of damage. Social science reveals that neither does it much correlate with education or income levels, landownership, species involved, personal losses to wildlife, or any other obvious characteristic. Work on wolves illustrates that tolerance for wildlife may be shaped by such mercurial constructs as attitudes, values, beliefs, social norms, culture, socio-political contexts, and media. HWC expert Alexandra Zimmermann concludes that the absence of ubiquitous drivers explains why solutions in one setting do not easily transfer to others (and may make things worse). New conceptual approaches are urgently needed to understand and address HWC, which is exactly what John Vucetich's team strove to provide within the non-anthropocentric framework elaborated in Chapter 3.

The root problem lies in incommensurables—things measured by different metrics that thereby thwart comparison—illustrated in microcosm by attitudes to a red fox: whether a particular red fox is a good or abhorrent thing would vary between the perceptions of the gamekeeper (villain that 'kills for fun'), photographer (beauteous subject), huntsman (noble quarry), one farmer (helpful slayer of pestilential rabbits), another farmer (predator of lambs), rights activist (victim of wanton cruelty), zealous veterinarian (purveyor of parasites), furrier (trim for a parka), while the naturalist-scientist sees a generally maligned, sometimes problematic, but always marvellously adaptable essential predatory cog in the functioning of the rural ecosystem . . . , and plenty more views, each held with conviction. The metrics with which each protagonist champions their opinion might be game-bird mortality, circulating stress hormones, capacity to suffer, lengths of farmland hedgerow not grubbed out, spinneys planted, lamb deaths, full-time equivalents of rural employment, revenue generated by tourism, and R_0, or some intangible metric of the value of tradition and cultural heritage, not to mention justice, freedom of speech, decision, and democracy. The agony of incommensurables reveals the multi-faceted fox as a bystander to conflict which rampages between people—a general truth about HWC is that it is largely between people, although consideration of the wildlife's interests is nonetheless relevant.

Solutions addressing symptoms not causes (often socio-political) may be skin deep, temporary, and counter-productive. Conceptual models of the intensity, intractability, and various approaches to resolution are based on the theoretical tenet that different levels of conflict exist and each must be addressed appropriately to their context and source. Francine Madden and Brian McQuinn propose a three-level categorization of deep-rooted differences underlying conflict to determine the approach and resources required to deliver resolution.

Three levels of conflict over wildlife

The symptoms of a Level 1—dispute level—conflict may present as a dispute over resources or safety. This could involve damage by wildlife to crops, livestock, fisheries, game, property, or facilities. Remember Bhutan: your tiger killed my cow. All HWCs start at Level 1, typically with tolerance of, and appreciation for, the problematic species, with affected communities neutral or cordial towards third parties. Practical solutions can settle the matter. These might be barriers, repellents, conditioning, guarding, or other physical interventions. Consider lion depredation on cattle from 14 villages adjacent to Hwange National Park, Zimbabwe: GPS tracking of cows and lions revealed that during the growing season, villagers intent on preventing their cattle eating their freshly planted crops herded livestock to woodlands far from villages but close to the Park and the lions, whilst during dry months cattle grazed close to villages where they were somewhat distanced from lions. The solution to ameliorate growing season risks involved communal herding, intensified guarding, and avoidance of densest woodland.

Level 2—underlying conflicts—arises where the visible issues of Level 1 occur against the bruised background of previous unsatisfactory solutions. Perceptions are distorted through the lens of anger and loss, with mounting resentment toward the species and meddling do-gooders. Accumulated disappointments or misunderstandings spawn a 'them versus us' mentality. Lessons on holistic approaches to avoiding this escalation are illustrated by two of WildCRU's coexistence projects: the Trans-Kalahari Predator Programme (TKPP) in Zimbabwe, and the Ruaha Conservation Project (RCP) in Tanzania—both dedicated to mitigating human–carnivore (especially lion) conflict in villages adjoining, respectively, Hwange and Ruaha National Parks.

Between 2008 and 2021 the Zimbabwean villages lost over 2,500 stock to lions, at an annual cost of US$49,000, directly affecting 14 per cent of households annually. Furthermore, a lion killed one child. Over 13 years 69 lions were killed due to conflict, with most killed on communal lands and fewest on hunting concessions. Consequently, prides at the core of the National Park were twice the size of those at the edge where perturbation resulted in only 42 per cent remaining intact for two years and only 48 per cent of their cubs surviving to one year. The holistic solution: TKPP recruited Long Shield Community Guardians, recommended by local traditional leaders, with good standing in the community and experience with predators. Following more than 1,200 cellphone messages warning farmers of approaching lions, and 75 occasions when Long Shields chased lions off with a cacophony of vuvuzela trumpeting, local manager Lovemore Sibanda's research demonstrated that the programme won local support, built relationships, and improved coexistence. Chasing lions away from cattle had immediate effect, and males that had been chased subsequently veered away from villages. Chasing was less successful in preventing re-offending the further it occurred from the park boundary and the more lions involved, the less stable their pride's membership, and in wetter seasons. The lowest rate of re-offending was amongst lions that experienced punishment on 40 per cent or more of the occasions they offended. Habitual re-offenders were the hardest to reform: lions that had killed more than 20 cows are almost beyond redemption, the worst killed 56. Of three reformed lions, two had been chased following their first infraction.

The villagers' food security was improved not only by reducing livestock predation but also by guiding farmers, who previously herded a handful of cows each, to form labour-saving cooperatives. Each cooperative was provided with easily erected, portable, lion-proof bomas within which up to 100 cows can be gathered at night, wherein they dung copiously. Maize grown on boma-treated ground grew bigger and more abundantly (Figure 20). Further, every local school received a comic book on the TKPP's work, in Ndebele and English, and the project's theatre group performed their coexistence narrative to more than 6,000 village children. Turning from carrot to stick, the project ran its own anti-poaching team. The TKPP is now scaling versions of these interventions throughout Zimbabwe and Botswana.

Paralleling the TKPP, the RCP has reduced livestock depredation by 60 per cent and reprisal killings of carnivores by 80 per cent. It too constructs bomas (safeguarding $2 million worth of stock for less than $50,000), alerts the community to approaching GPS-tagged lions, and imports stock-guarding dogs. The RCP has developed a portfolio of community benefits: 15 local primary schools are twinned with UK and US schools (each raises £300 per year for books, pens, desks), more than 40 Simba Scholars have been funded at secondary school, and every day it provides 800 school breakfasts (mothers react angrily when these are jeopardized by community members breaking the rules of coexistence with wildlife), and has equipped a clinic. The key innovation is to reward coexistence (the principle of Payments for Ecosystem Services, elaborated in Chapter 9): farmers making their bomas predator-proof are given grants for veterinary medicines—a concept familiar to English farmers

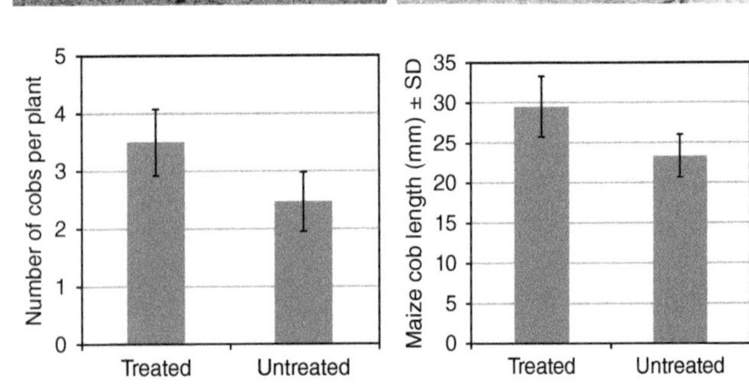

20. The ground inside (treated) and outside (untreated) the bomas, and the resulting difference in number of maize cobs and cob length.

who, before Brexit, received £427 million p.a. of payments under agri-environment schemes. While English farmers may get payments for producing skylarks, Ruaha's pastoralists are given camera-traps and awarded points for the wildlife they evidence on their beat. A young warrior who might formerly have gained recognition and rewards by spearing lions is instead rewarded for getting a photo of them: a photo of a small antelope scores 1,000 points, baboon 1,500 points, spotted hyaena 10,000, lion 15,000. Not in my backyard? Actually, yes please, when a wild dog is worth 20,000 points, and a pack of them can deliver a bonanza of

340,000, and the monthly tally of points is converted into US$. Wildlife generates employment, and points generate health, education, and veterinary benefits—more benefits than any other local initiative.

HWC escalates to Level 3—identity-based conflict—if underlying issues align with pre-existing socio-political conflicts, so that stakeholders perceive HWC (and those trying to solve it) as threatening their values or communal identity. The resulting impasse is operationally difficult and ethically challenging. It is as unsurprising as it is shocking that while I wrote this chapter a Guardian newspaper headline reported, 'Wolves shot in Norway after court overturns stay of execution'. In parts of Europe and in North America, the Little Red Riding Hood legacy leaves wolves the subject of intractable human–wildlife conflicts. Put coarsely, some people hate wolves and other people don't. Norwegian and French farmers suspect their governments of secretly reintroducing wolves, while in Slovakia deep hatred of wolves seethes despite negligible depredation on sheep, including a 70 per cent reduction thanks to Slovenski Cuvac stock-guarding dogs. At Level 3, financial compensation is insufficient, as the 'problem animal' becomes a totem of political, economic, and cultural conflict—a lesson relevant also to reintroducing predators such as lynx to Britain.

Diagnosing levels of conflict

Conflict Resolution Theory, developed from international relations and peacekeeping, warns of mismatch between levels of problem and solution. This mismatch risks widening the rift between those affected by the problem and those seeking to resolve it. General advice to those volunteering a voice for wildlife is to strive to 'do no harm' while negotiating the operational and ethical quagmire: to avoid mismatch, tread warily. A first step is to understand the values and beliefs of the stakeholders, through surveys, participant observation, focus groups, and a plethora of increasingly sophisticated techniques absorbed from the social sciences. The Zimmermann 2020 Conflict Framework (ZCF) guides practitioners in asking the right questions across five categories: (a) perceptions about the species blamed in the conflict, (b) questions about the apparent situation itself, (c) questions about the history of attempts to address the conflict, (d) the extent of willingness to find solutions, and (e) views about others involved in, or trying to assist with, the situation.

As to diagnosis, at Level 1, one might expect some empathy for the wildlife in conflict, with concern focused on tangible impacts or losses, and a perception that attempted resolution was helpful. Giant armadillos feed mainly on ants and termites, but in Mato Grosso cause substantial cost to beekeepers by using their 60 kg bulk to topple beehives, to which they return on successive nights, exposing themselves to retributive poisoning. Using the ZCF, conservationists diagnosed Level 1 conflict, soluble operationally (by raised hives and electric fences), and turned to advantage by a beekeeper certification scheme. At Level 2 parties express accumulated frustration, perceive a major problem, exaggerate its frequency and impact, and resent third parties. Level 3 is characterized by strongly negative, exaggerated views on the species, and perhaps wildlife in general, and wide apportioning of blame: conservationists hearing reference to 'your' animal can expect trouble. Protagonists are likely to be unbending and dismissive of previous attempted solutions. So, while Level 1's focus on negotiable interests can be approached through compromise and practicality, Level 2 necessitates delving into history to build relationships on common ground, while Level 3 requires reconciliation dialogues acknowledging that suggesting compromise of a fundamental value can feel like an assault to dignity. Conflict about wildlife can require multi-track diplomacy and the vocabulary of power imbalances and redistributed decision-making: solutions are likely to lie in facilitated dialogue including deliberative processes, remembering Herbert Spencer's remark that 'opinion is ultimately determined by the feelings not the intellect'.

Sometimes there's consensus: Scottish creel fishing for lobsters and crabs involves dropping baited baskets—creels—from boat to seabed on lines, which can cruelly, and fatally, entangle humpback whales, dolphins, porpoises, and basking sharks. The economic impact on fishing is negligible, but entanglement may impact conservation and certainly creates a welfare horror as distressing to fishers as to anybody else. A shared quest for solutions to safeguard marine biodiversity and traditional fisheries is driving industry-led research. Technical fixes are as numerous as the nuances of HWC are legion (Figure 21). The conflict and mitigations become selective pressure on the target species.

An iterative adaptive management approach (Figure 22) progresses towards solving conflicts by whittling away at problems. This applies well to the rampant conflicts with wild canids that embody culture wars between different human world-views. Consider a game farm in the South African Karoo, where black-backed jackals got 50–80 per cent of their diet from the fawns of antelope, costing the game farmer about $10,000 a year, whereas on a sheep farm jackal diet was 30–50 per cent sheep, and they typically killed 100 lambs annually, costing about $20,000. The problem is potentially reducible on the game farm by swapping antelope species that 'hide' their newborn for species where the newborn stay with, and are protected by, their mother from jackal predation. The sheep farmer might reduce the problem by shifting the lambing season to coincide with the peak in wild ungulate births, increasing the jackals' opportunities to take wild prey, and the sheep could be protected by non-lethal methods such as guard dogs. In both cases, an irreducible component will remain, which may or may not be bearable.

Culture, psychology, and beyond

Any reader who innocently thinks conservation is still the prerogative of backwoodsmen naturalists reading tracks in the mud should dip a toe in the rather hotter water of the higher-level

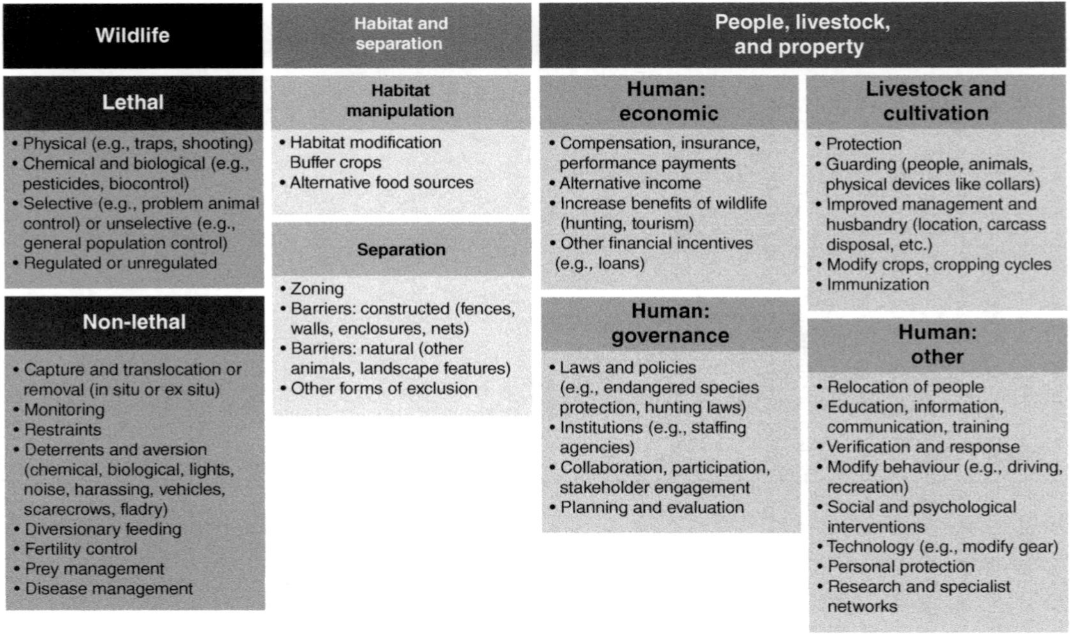

Wildlife

Lethal

- Physical (e.g., traps, shooting)
- Chemical and biological (e.g., pesticides, biocontrol)
- Selective (e.g., problem animal control) or unselective (e.g., general population control)
- Regulated or unregulated

Non-lethal

- Capture and translocation or removal (in situ or ex situ)
- Monitoring
- Restraints
- Deterrents and aversion (chemical, biological, lights, noise, harassing, vehicles, scarecrows, fladry)
- Diversionary feeding
- Fertility control
- Prey management
- Disease management

Habitat and separation

Habitat manipulation

- Habitat modification
  Buffer crops
- Alternative food sources

Separation

- Zoning
- Barriers: constructed (fences, walls, enclosures, nets)
- Barriers: natural (other animals, landscape features)
- Other forms of exclusion

People, livestock, and property

Human: economic

- Compensation, insurance, performance payments
- Alternative income
- Increase benefits of wildlife (hunting, tourism)
- Other financial incentives (e.g., loans)

Human: governance

- Laws and policies (e.g., endangered species protection, hunting laws)
- Institutions (e.g., staffing agencies)
- Collaboration, participation, stakeholder engagement
- Planning and evaluation

Livestock and cultivation

- Protection
- Guarding (people, animals, physical devices like collars)
- Improved management and husbandry (location, carcass disposal, etc.)
- Modify crops, cropping cycles
- Immunization

Human: other

- Relocation of people
- Education, information, communication, training
- Verification and response
- Modify behaviour (e.g., driving, recreation)
- Social and psychological interventions
- Technology (e.g., modify gear)
- Personal protection
- Research and specialist networks

21. Common approaches used to mitigate human–wildlife conflict and promote human–wildlife coexistence.

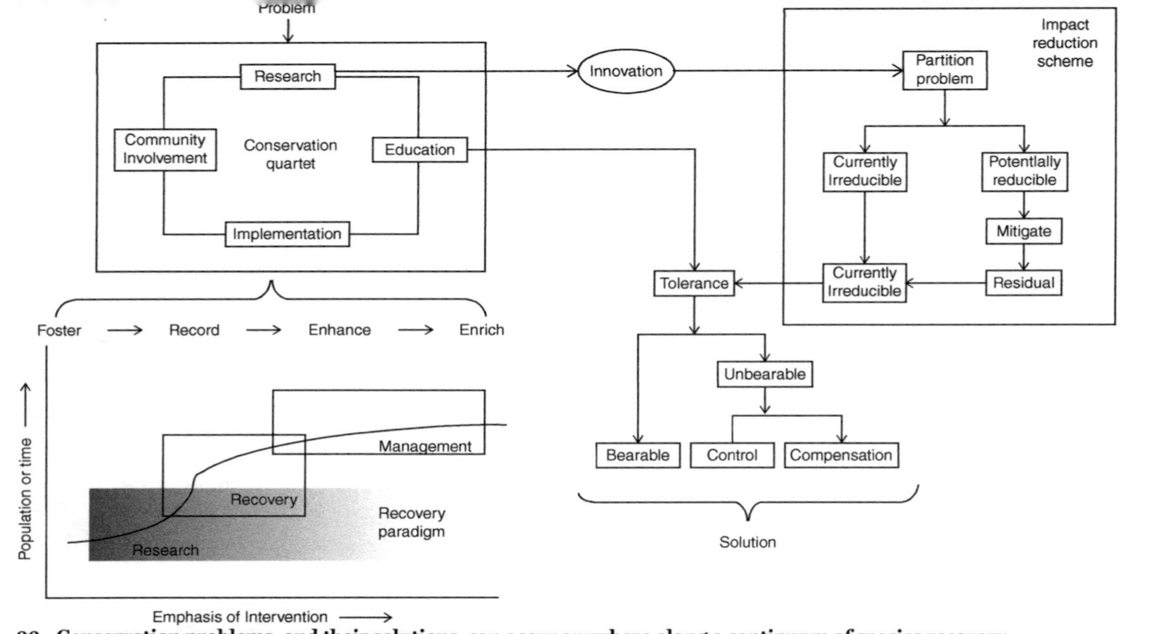

22. Conservation problems, and their solutions, can occur anywhere along a continuum of species recovery, developing the notion that problems can be partitioned between reducible and irreducible elements, and the balance between these will shift as currently intractable elements are rendered reducible by new innovation.

holism that leads to Conservation Geopolitics (Chapter 10). A quick insight into how different is the modern conservation landscape is given by Robert Fletcher and Svetoslava Toncheva in 'The political economy of coexistence', and its argument that a key to understanding HWC is 'the global capitalist political economy and the uneven geographical development it produces'. They illustrate this by the contrasting political economics of jaguars in neoliberal Costa Rica (only 10–20 survive, crushed hopelessly between farmers and international investors), versus brown bears in post-socialist Bulgaria (600–800 survive and motivate the tourism that contributes to the employment of some 90 per cent of people in the studied village).

There's no dodging the pithy conclusion of Amy Dickman and her colleagues that 'humans are the common thread in the highly variable arena of human–wildlife conflict, and the course and resolution of conflict are determined by the thoughts and actions of the people involved, understanding the human dimension is the most crucial prerequisite for developing effective mitigation'. Culture is key: three communities of African cattle farmers, facing the same problem—lions killing their cows—reacted very differently. In Kenya's Maasailand, 88 per cent of the respondents expressed a desire to see current lion populations maintained, while the figure was only 42 per cent in Tanzania's Ruaha and only 5 per cent of those in Zimbabwe's Hwange landscape. The perception of personal benefits from conservation was greatest in Maasailand, although exposure to conservation education was highest in Ruaha. Statistical techniques revealed that combining personal benefits with conservation education was the most probable route to increasing an individual's desire to see current lion populations maintained. The importance of culture is further illustrated by attitudes to the Asiatic lion in India's Greater Gir landscape. Conservation there delivered more lions, but so too came more stock losses. Farmers suffered crop damage by wild ungulates, and believed that lions killed, or scared off, deer and other crop pests. Remarkably, 77 per cent of respondents were positive about lions—their tolerance was largely cultural. Unexpectedly, only 28 per cent were positive about leopards, although leopards made only a fifth as many attacks on livestock.

Scholars of HWC increasingly turn to psychology. Of the myriad factors that might have determined the care African stockmen took to protect their livestock from predators, social norms emerged as predominant: in short, if your friends and neighbours take good care of their cows, you're likely to do so too, and vice versa.

Following 17 studies across seven countries in Central and South America, Zimmermann's team searched in vain for socio-economic commonalities of human–jaguar conflict. Within and across case studies there were differences in farmers' education, economic dependence on livestock, personal experience with livestock losses, as well as tolerance of and attitudes and social norms towards jaguars. However, no single quantifiable contextual factor could predict how farmers perceived jaguars and dealt with their depredation. Patterns in HWC inform action only at a local scale, and even if a given solution begins to look like a silver bullet for a handful of cases it would be unwise to assume it transferable. Further, the same species can get into different sorts of trouble in different places. In China's largest tropical forest, in Xishuangbanna, while land was converted to rubber plantations, Asian elephant numbers and HWC have increased. In Cox's Bazar district of Bangladesh 930,000 Rohingya refugees arrived from Myanmar and built camps in a movement corridor of critically endangered wild Asian elephants. They entered into camps, destroying settlements, killing 13 refugees, and injuring 50 more.

HWC will generally be worsened by climate change—as illustrated by the elephants with less natural forage available due to the heatwave in India in 1986–8. The elephants entered villages' lands, raiding crops and killing people. The rate of polar bear attacks has trebled in Canada's Hudson Bay during the last 30 years of melting sea ice, and between 2010 and 2014, when sea ice extent reached record lows, there were 15 attacks, the greatest number ever recorded in a four-year period, whilst in 2021 eastern Australia saw crops destroyed, grain silos and barns infested, and homes invaded by a massive plague of mice following a three-year drought.

Implications for practice and policy

Emphasis in tackling HWC should shift from reactive crisis management to proactive anticipation that ameliorates underlying issues and thus prevents Level 1 conflicts worsening. Simon Pooley's review advocates structured understanding of conflict before it becomes entrenched, fostering a do-no-harm approach, transdisciplinarity, and holistic thinking. Another leader in this field, Silvio Marchini, emphasizes that attention to levels of conflict must be incorporated early in any theory of change in conservation planning, whether by non-profits, institutions, or governments. These ideas permeated the IUCN Human Wildlife Conflict and Coexistence Specialist Groups' 2023 guidelines (https://www.hwctf.org/guidelines).

Chapter 8
Climate change

Picture an Ethiopian wolf, its fur fiery red, its elongated, rodent-thumping muzzle pointing elegantly upward as it surveys what further refuge lies above the plateau of its last mountain-top sanctuary. It is already one of only 500 or so survivors of the rarest canid species in the world, a long-term victim of some disadvantageous throws of the Pleistocene dice that fragmented its Afro-alpine habitat. More recently, these rarest of wolves have been buffeted by a succession of conservation detriments—conflict with shepherds, rabies, hybridization with domestic dogs, target practice by insurgents. The last thing this species needed was a further straw to break its already burdened back. But that final straw now grows weightier as the wolf's upward glance reveals no uphill refuge—there is none: it is already at the top; above lies only the thin air of extinction. This plight, nowhere left to go, is painfully familiar in images of polar bears adrift on shrinking ice floes bereft of seals, and less conspicuously played out for countless unseen species whose shrinking distributional envelopes approach implosion. Consider Atlantic cod, that thrive in cold, deep sea regions whose decline from overfishing over the last century was worsened by rising ocean temperatures. Or the Kalahari pangolin, whose supply of ants has, literally, dried up. These are just some of the individual tragedies, the dislocated moving parts of ecosystems whose processes strain perilously as climate changes. Direct measurements of CO_2 in the atmosphere, and in air extracted from ice cores, show that atmospheric CO_2 has increased by about 45 per cent from 1800 to 2018, with current concentrations the highest they have been for 800,000 years. These changes have occurred because the balance of the carbon cycle has been disturbed by the additional CO_2 released by man-made activities such as burning fossil fuels, deforestation, and farming.

An earlier generation of conservation biologists appreciated Jared Diamond's memorable 1989 taxonomy—the 'Evil Quartet'—of the challenges biodiversity faced: habitat loss/ fragmentation, overkill, introduced species, and extinction chains. Each of these challenges is worsened by changing climate. The pervasiveness of the impacts is captured by the World Wildlife Fund for Nature:

> Sea levels are rising and oceans are becoming warmer. Longer, more intense droughts threaten crops, wildlife and freshwater supplies. From polar bears in the Arctic to marine turtles off the coast of Africa, our planet's diversity of life is at risk from the changing climate.

Writing now, in the aftermath of COP26 (United Nations Climate Change Conference UK 2021) in Glasgow, it is hard to recall that scarcely a decade ago the educated general public—even a lecture theatre full of undergraduates taking an option on biodiversity conservation—needed to be convinced that the evidence for climate change was real. Nowadays, detail about climate change is familiar to everyone, and thanks to young activists such as Greta Thunberg, most schoolchildren know more detail today than most professors did a decade ago. Professionals agree that biodiversity, and thus humanity, faces disaster: in 2022 upward of 14,000 scientists had signed the World Scientists' Warning of a Climate Emergency, and the threat to biodiversity is so severe that in 2021 conservation scientists Charlie Gardner and James Bullock called for 'a paradigm shift from biodiversity conservation to survival ecology' and wrote forcefully that 'earth faces a climate emergency which renders conservation goals largely obsolete'.

One aspect of climate change, global warming, is characterized by the now famous 'hockey stick plot': a steep rise in temperatures since the 1850s. Sceptics emphasize that a bit less than 200 years of data are scarcely a blip in the earth's history—yes, on a geological time-scale the earth today is not especially hot. In a palaeontological context CO_2 levels are not high, and rapid rates of climate change are not unprecedented. In the theatre of evolution, climate change has always been a major driver of adaptation. These truths do not detract from the fact that climate change caused by humans is unique, and adds woefully to the dreadful litany of anthropogenic threats to biodiversity. Things are, anyway, different now, with the pervasive frailty of species and ecosystems due to a myriad man-made impacts, and a mass extinction already at hand. The year 2021 saw catastrophic wildfires rage in California, Greek islands evacuated, and temperatures in Sicily reach a historical record of 48.8°C. The biological problem is that humans—and all other animals—find unpredictable and uncontrollable change or extremes stressful, even unmanageable, beyond the behavioural and physiological adaptability of their species.

The average global temperature on Earth has increased by a little more than 1°C since 1880. Two-thirds of that warming has occurred since 1975, roughly 0.15–0.20°C per decade. The rise in temperature is not uniform worldwide: all regions are warming, but some, such as polar ice-caps, are more affected. Receding polar ice means that less sunlight gets reflected into space and is instead absorbed into the oceans and land, raising the overall temperature, and fuelling further melting. The size of the Arctic ice-cap has decreased by over 40 per cent since the late 1970s, particularly in summer and autumn. If the entire ice-cap melted, global sea level would rise by 7 m, and if the West Antarctic ice sheet melted, the rise would be 3 m. Global average sea level

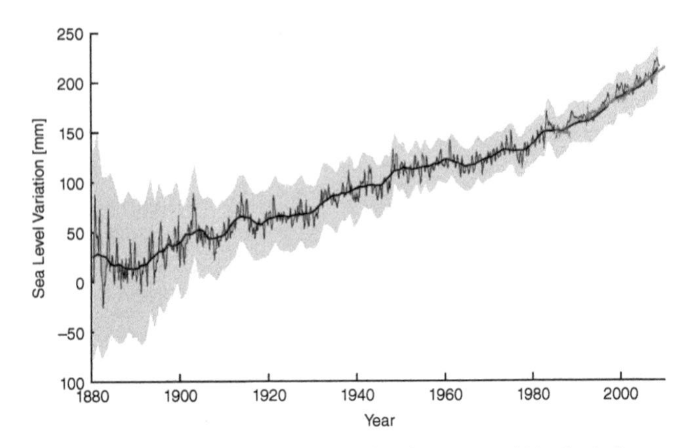

23. Global average sea level has risen by about 16 cm (6 inches) since the late 19th century, and faster recently; tide gauges (black line) and satellites (jagged line) indicate that average sea level rise over the last decade centres on 3.6 mm per year. Shading represents uncertainty, which has decreased as the number of global gauge sites data points have increased.

(Figure 23) has risen by about 20 cm since the late 19th century, and the rate is accelerating: over the last two decades it has risen 3.2 mm per year. Bangladesh, with the world's largest alluvial delta, could see 17 per cent of its land mass disappear with a 1 m rise in sea level, submerging the shrimp-farming livelihood of 150 million inhabitants, and spelling the end for the tigers of the Sundarbans. On dry land, desertification rates are increasing, precipitation patterns changing, and agricultural yields are adrift. Conservation strategies under climate change need to be adaptive, participatory, and transformative.

Biological effects

Weather—a short-term expression of climate—is a fundamental dimension of the niches of animals and plants, and so it is obvious that most organisms would be impacted by climate change. Against the 'normal' or 'optimal' range of weather that characterizes a species' bioclimatic niche, variability is how extreme (and/or rapid) fluctuations are, and unpredictability is how unusual. Cascading from impacts on individuals—the baleful upward glance of that Ethiopian wolf—the consequences for populations include changing distributions and ranges, increasing extinctions, altering population dynamics, phenological (i.e. natural timings) upset, and increasing likelihood of disease and of invasive species. I first came across some principles shaping such a cascade in the late 1980s, before the term climate change was popularized, while working with my then doctoral student, and subsequent doyen of Icelandic ecology, Pall Hersteinsson. We considered then-emerging changes in the geographical distributions of red and arctic foxes with largely separate geographical ranges that overlapped in the tundras of North America and Eurasia. The overlap was partly the result of changes in the distribution of red foxes during the early 1900s, during which decades they expanded the northern limits of their distribution into higher latitudes and altitudes. As the behaviour and biology of arctic and red foxes are similar, interspecific competition (between different species) occurs where they meet. Indeed, decreases in arctic fox distribution and abundance mirror, in reverse, the red fox's range expansions. We focused on the energetic consequences of the difference in body size between these two otherwise very similar species—the red fox's larger size simply makes it too expensive to operate in environments with low primary and secondary productivity, such as the arctic tundra. Historical records confirmed that arctic fox pelts formed a lower percentage of fur harvests as mean July temperatures increased. The uphill creep of red foxes in Sweden increased their altitudinal range by 330 m per 2°C rise in temperature, pushing the arctic foxes to precarious heights and diminishing areas above them.

Until 1940 European badgers occurred no further north than the scattered populations at ~63° N in southern Scandinavia. Just 40 years later badgers occurred 300 km northward, beyond the Arctic Circle, up to 67° N in Sweden and 68° N in Norway. By

1995 they were settled along the Baltic coast in Sweden and inland to 65° N in Norway. In Finland they have spread northward by 100 km since 1940 and their numbers in the south have doubled. The decimation of wolves by people may have opened doors for these badgers, but the increasing length of the snow-free season and its average temperature has been key.

A synthesis of physiological and behavioural data evaluated the costs of chronic exposure to sustained hot weather, caused by ongoing warming, for birds inhabiting southern Africa's Kalahari Desert over the course of the 21st century. The sub-lethal costs of chronic exposure, manifested as progressive loss of body condition, delayed fledging, reduced fledging size, and outright breeding failure, will most likely drive major population declines, leading to the loss of much of the Kalahari's avian biodiversity before the next century.

Rising ocean temperatures kill algae, causing many Galápagos-dwelling marine iguana to starve, whilst warmer weather can interfere with their beach-nesting, egg development, and ability to regulate body temperature. The migratory Bluefin tuna feeds in cold North Atlantic oceans, but heads to tropical waters to reproduce; increasing ocean temperatures have negatively affected this endangered tuna species' mating and reproductive habits, particularly in the Gulf of Mexico. Indeed the detrimental effects of global warming span the largest animal on Earth, the blue whale, where melting Antarctic ice is diminishing krill populations, the blue whale's main food source, to the smallest such as insects—despite successfully diversifying through more than 450 million years of Earth's changeable climate, many insects are finding the speed with which the climate is changing difficult to adapt to. Rusty patched bees were once prevalent in US grasslands and tall grass prairies, but high temperatures and extreme rain, as well as drought and premature snow melting, hinder the flowering of pollinator plants, restrict areas for queens to nest and hibernate, and shorten the time for honey-gathering, forcing the species to the brink of extinction. Possibly of greatest concern are the consequences that climate change has for plants, be it heatwaves, increased flooding, droughts, or the way rising carbon dioxide concentrations and temperatures directly affect plant growth, reproduction, and resilience. The impacts on natural vegetation and agriculture are difficult to predict but could have devastating consequences for humankind.

Another weather-related phenomenon is El Niño, which describes the warming of sea surface temperature that occurs every few years, typically concentrated in the central-east equatorial Pacific. As recently as the millennium, when Ecuadorian conservationist Hernan Vargas began research on Galápagos penguins, it was not clear whether the apparent increase in frequency and severity of El Niños was just bad luck—now it is accepted as a symptom of climate change. Fish stocks decline in response to El Niños, which are detrimental for all seabirds, but particularly so for the Galápagos penguins because of their highly localized distribution. Following extreme El Niño events in 1982–3 and 1997–8 the penguins' populations declined by more than 60 per cent. El Niños not only cut off penguins' food supply under water, but also drenched them with chilling rain as they shivered on their nests—there are currently fewer than 600 breeding pairs.

Meanwhile, on shore, the endemic rice rat on the Galápagos island of Santiago teeters towards extinction and this is likely to worsen if El Niño events become more frequent. Recall that these rare rice rats are already threatened by invasive ship rats as well as climate change. The larger ship rats are aggressively dominant but suffer periodic population crashes when struck by La Niña droughts on the arid north coast of Santiago. As mentioned in Chapter 4, this offers the endemic and arid-adapted rice rat sufficient respite to recover because it can survive on a specialist diet of *Opuntia* cacti, which the invader cannot digest. However, the rain-drenched El Niño years allow lush vegetation to flourish and thus enable ship rats to recover, intensifying interference competition between the rodent species— competition in which the endemic rice rat loses to the invasive ship rat.

Climate change can also cause subtle changes in species populations, demonstrated especially well by the gold-dust of Ecology: long-term data, which turn a snapshot into a documentary. This can reveal phenological disengagement: when processes that have evolved in synchrony fall out of tune. Fifty years of data in Wytham Woods, Oxfordshire, revealed that warmer springs have led to a three-week advance in the great tits' laying date; the timing of their key food resource, caterpillars, has advanced at the same rate. So far, then, these cogs in the ecosystem remain engaged, and both predator and prey are keeping pace with the warming. Meandering by night beneath the great tits, the badgers of Wytham Woods, also monitored closely for 50 years, illustrate further complexity. They feed mainly on earthworms, which only surface on mild, moist nights. Between 1987 and 1996 badger numbers more than doubled: 64 per cent of this increase was driven by adult survival, 30 per cent by juvenile survival, and 5 per cent by litter size, coinciding with unusually warm years in the late 1990s. Badgers grew significantly heavier in milder Januaries, leading to greater overwinter survival and more litters born. However, their greater activity in warmer winters leads them into more road traffic accidents and dry springs cause greater parasite loads. Analysis of the 'Goldilocks zone' (neither too hot nor too cold—the expected weather to which they are adapted) demonstrated that temperature, rainfall, and the amplitude of change between winter and summer explained changes in badger populations.

The pervasive risk of mismatch threatens ecosystem processes: migratory brant geese experience changing triggers to migrate to and from Alaska—arriving there earlier they reduce plant biomass, which in turn alters the flux of nitrogen and carbon through the grassland, changing a summer carbon sink to a source. Remarkable research in 2020 led by Christopher Trisos modelled the consequences to 30,000 marine and terrestrial species of finding themselves in unprecedented temperatures. Because many might share similar tolerances, and hence tip into trouble at the same point, such 'near simultaneous exposure among multiple species could have sudden and devastating effects on local biodiversity and ecosystem services'. That tipping point might occur when 20 per cent of species in an assemblage undergo exposure to unprecedented temperatures within the same decade. The researchers foresee imminent radical disruptions to biodiversity: under high emissions scenarios, in tropical oceans by 2030, and reaching tropical forests and higher latitudes by 2050. The geographical position of this temperature niche is likely to shift more in the next 50 years than in the previous 6,000, putting 1–3 billion people, and their agricultural systems, into extreme discomfort, if not societal collapse.

During the early part of the 21st century, an unprecedented change in the status of vector-borne diseases in Europe has occurred as species have shifted their ranges. For example, malaria has re-emerged in Greece, and West Nile virus has reached parts of eastern Europe. Since 1990, five different species of alien *Aedes* mosquitoes have become established in Europe. Amongst them, the Asian tiger mosquito can transmit infectious diseases such as dengue and chikungunya fever. Projections are, unusually, unanimous: the distribution of tiger mosquitoes will shift to northern countries, including the UK, over the next decades, when wetter and warmer conditions will allow the mosquito to overwinter in the north.

Biodiversity drives the cycles that allow life on Earth and the feedback loops that stabilize climate (via CO_2), prompting the expert in global sustainability Johan Rockstrom in 2021 to call for 'biosphere stewardship that protects carbon sinks and builds resilience'. The extinction of mammoths, and consequent loss of open ground (and thus reflective surface or albedo) to shrub-cover, may have increased the temperature in Siberia by 1°C. Searching for alignment between large mammal conservation and climate change mitigation, Yadvinder Malhi's team concluded in 2022 that synergy existed through large mammals causing beneficial changes in fire regimes in dry grasslands, terrestrial albedo (herbivores causing shifts from closed to open canopy-cover at high latitudes), and carbon stocks in grassland vegetation and soil—all especially in tundra biomes. What matters is not the carbon tied up in mammals globally—'[t]he total carbon in wild mammals and birds is equivalent to roughly eight hours of current anthropogenic fossil fuel emissions'—but their scalable effects. These contribute to climate change adaptation (helping biomes exhibit resistance, resilience, and transformation) through ecosystem intricacy. For example, large herbivores in tropical forests can increase above-ground tree biomass by 60 tonnes/ha by reducing competition from juvenile trees and favouring dispersal of large-seeded (high-density wood) trees.

The future

Traditionally, nature conservation has largely focused on maintaining the status quo of biodiversity. Do things need to change in the face of climate change? Should policy-makers opt for hard sea defences or allow the increase of coastal salt marshes? And should they assist species to move to places where they might survive? In England the government's statutory body for conservation tends to favour 'adaptive management', albeit under the limiting constraints of too few resources and too much uncertainty. In the USA the legal mandate to protect 'the community of life', particularly in wilderness, allows the US Forest Service to protect entire ecosystems and the wide range of environmental gradients necessary for species migration, dispersal, and viable populations as the climate changes. In the context of biodiversity conservation policy-makers almost always face a multiplicity of problems and paucity of funds and capacity, which puts a premium on efficiency and the quest for win-wins.

Forests are a pre-eminent nature-based solution to the climate emergency. Tropical forest biomass contains an estimated 323 billion tonnes of carbon and two-thirds of the world's terrestrial biodiversity. If deforestation is stopped and damaged forests are restored, this could provide around a third of the carbon reduction needed. The UN's REDD (Reducing Emissions from Deforestation and Degradation) mechanism proposes multi-million-dollar payments from developed countries to developing nations in return for improved forest management, and the consequent reduction in carbon dioxide emissions. The addition of a biodiversity component to 'REDD+' may offer mechanisms to conserve key species. An attempt to combine conservation and carbon in Brazil led to the idea of jaguar 'REDDspots', defined as the municipalities that offer the best opportunities for co-benefits between the conservation of forest carbon stocks, jaguars, and other wildlife, identifying win-wins in 25 per cent of 40 municipalities. A further 95 municipalities had potential to develop additional REDD+ projects, offering great rewards for biodiversity conservation in Amazonian and Atlantic Forest biomes.

I remember Margaret Thatcher's 1989 speech to the UN—arguably the first acknowledgement of climate change by a world leader. Subsequently, carbon in the atmosphere has increased relentlessly by 2 ppm p.a., so whatever has been done hasn't worked: we've burnt ever more coal and destroyed ever more of biodiversity's capacity to sequestrate (Brazil currently destroys Amazonian forest at 1 ha/min). In 2020 the UK Climate Change Committee reported 'by reducing emissions from the UK to zero we also end our contributions to rising global temperatures'. There is plenty of smoke on this political mirror, because the UK has actually consumed even more carbon-heavy goods, but exported their production, mainly to China, to support our unsustainable carbon lifestyles. Following the lacklustre outcomes of Kyoto (1997) and Paris (2015), the core outcome of COP26 negotiations in 2022 was a climate deal agreed by 197 countries, the Glasgow Climate Pact.

Unlike its predecessors, the 2022 pact recognized the connections between the climate and nature agendas, and the critical role of protecting, conserving, and restoring ecosystems to deliver benefits for both climate adaptation and mitigation. COP26 also saw major nature-related announcements, with a strong focus on halting deforestation and shifting to more sustainable farming practices. Over 140 countries, accounting for more than 90 per cent of the world's forest, committed to end deforestation by 2030. Forty-five governments, representing 75 per cent of global trade, released the Financial Sector Roadmap for Eliminating Commodity Driven Deforestation. And commitments from the public and private sector amounted to approximately US$19 billion for nature and land-use. Announcements at earlier COPs did not prevent deforestation rates reaching an all-time high in 2021, and carbon levels rising unabated—so what is different this time? There is more money, more pressure from civil society to hold businesses and governments to account, a clearer understanding of the risks and opportunities associated with biodiversity loss, and businesses may be ready to engage.

However, emissions will reduce only with change from carbon-based agriculture, transport, and electricity-generating systems. Soil contains four times the carbon in the atmosphere: industrializing agriculture over the past century has contributed most to rising atmospheric carbon, both directly and by enabling (together with medical advances) the rise in the world's population from 2 to 8 billion. However well intentioned policy announcements by Western governments, it is China, India, and Africa (their GDPs doubling every decade) that will determine the bulk of future carbon emissions. Carbon emissions are driven by carbon consumption, which is commissioned to greater or lesser extents, at home and abroad, by us all. Carbon consumption is grossly inequitable between poor and affluent countries, a disparity facilitated by REDD+ schemes.

Even if humankind collectively manages to reverse global warming before species are irreversibly lost from ecosystems, the ecological disruption caused by raised temperatures could persist for half a century or more; a recent study that examined more than 30,000 species of mammals, birds, amphibians, reptiles, and marine fish and invertebrates found that for over a quarter of the worldwide locations studied, the chances of returning to the pre-overshoot 'normal' are uncertain or non-existent.

Of all the challenges facing biodiversity conservation, none makes more plain than the urgency of mitigating climate change that everybody's behaviour must change. This was acknowledged in compelling speeches at COP26, and echoed through a globally engaged public. However, the past has been far from promising, and as Dieter Helm pithily observes in his compelling book *Net Zero*, 'Failure to act does not abolish the consequences of not acting—they cannot be escaped.' The clear and present priorities are to reduce carbon emissions *and* consumption, develop alternative ways to generate the electricity that powers the digital age, and slow human population increase; some of these actions may, as a co-benefit, conserve biodiversity. However, it would be naive to assume that solving climate change for humanity, while urgently necessary, will be sufficient to deliver biodiversity to safety. Indeed, as climate impacts accelerate, it is urgent to reverse a sinister race to the brink, which risks the apocalyptic horsemen astride habitat degradation, persecution, invasives, and over-use overwhelming biodiversity before climate change delivers the *coup de grâce*.

Part 3

The way ahead

Chapter 9
Who pays, and how?

In 2021 the World Economic Forum identified five top global threats to humankind: extreme weather, climate action failure, human environmental damage, infectious disease, and biodiversity loss. The latter bumps into Economics from two directions. First, the human enterprise is costing the earth: our consumption far exceeds nature's capacity to supply. What economic instruments can reduce humanity's impact on biodiversity? Second, conserving biodiversity is often expensive, so how can it be paid for?

In 'The Economics of Biodiversity' (2021), Sir Partha Dasgupta argues that the solution starts with accepting that our economies are embedded within nature, not external to it. For economic growth to be sustainable, humanity's engagement with nature must be sustainable, thereby enhancing our collective wealth and well-being and those of our descendants. Business-as-usual is a doomed strategy—the Living Planet Index (a measure of the state of global biological diversity based on worldwide population trends of vertebrates) is declining at an accelerating rate; on current trends it will be at 15 per cent of its 1970 level by 2050. The Dasgupta proposition is that three changes are essential. The first is so obvious that he must have despaired at having to spell it out:

1. *Humanity's demands on nature must not exceed its supply.* Demand for food and fossil energy are major drivers of biodiversity loss. Therefore, advances in food production and decarbonization can ameliorate the impacts of climate, land-use, and oceanic degradation on biodiversity, rebalancing supply and demand. Conserving and restoring natural assets will sustain and enhance their ability to supply—prevention is better than cure: it is less costly to conserve nature than to restore it. Growing human populations, and their footprints, will determine patterns of global consumption. The developed world has the largest current impacts on global sustainability, yet most biodiversity and emerging threats are in the Global South which is moving towards similar goals, a friction that needs to be more openly addressed.
2. *Change measures of economic success to a more sustainable path.* Gross Domestic Product (GDP) is a widely used, but unsatisfactory, measure of economic success. GDP ignores the depreciation of the natural environment. It thus encourages the pursuit of unsustainable growth and development. Better to measure 'inclusive wealth' in terms of all assets, including natural capital (the world's stock of natural resources including geology, soils, air, water, and biodiversity).
3. *Transform finance and education systems to enable these changes and sustain them for future generations.* Biomes, and the ecosystems on which humanity relies, need to be paid for. There should be charges for the use of biomes (such as oceans) beyond national boundaries (and prohibitions where they are too ecologically sensitive to be used at all). The rich must make payments to help the poor. The financial system must channel investments towards economic activities that enhance natural assets and encourage sustainable consumption and production. Far from the muddy boots and binoculars that once characterized those striving to deliver biodiversity conservation, this will require the commitment of governments, central banks, and international financial institutions. It is estimated that to globally fund nature

protection would cost $100 billion a year, which is not so much considering that harmful public subsidies to agriculture and fossil fuels alone amount to $500 billion p.a.

One perilous error, recognized by economists since Pigou in 1920, but persistently overlooked in the pricing of almost everything, has been to ignore the true cost. That true cost of almost everything includes the impact on biodiversity and the wider environment. In short, despite recognizing externalities (where the action of, say, a polluter who does not pay, impacts another—often, all of us), the error has been not building them into pricing. The folly of false pricing—ignoring the cost of an unstable climate, polluted air, a thinned ozone layer, acidified oceans, and damaged biodiversity—prompted Johan Rockstrom's powerful concept of planetary boundaries and the 'safe operating space for humanity' (these boundaries have long ago been massively overshot for biodiversity loss). This space is accessibly depicted in Kate Raworth's *Doughnut Economics*, combining a social and an ecological boundary that together encompass humanity's well-being (Figure 24). The inner boundary represents the social foundation, beneath which lie deficits in hunger, ill health, illiteracy, whilst the outer boundary represents an ecological ceiling, beyond which lies deterioration of Earth's life-supporting systems built on biodiversity. Between these two boundaries lies ecologically safe and socially just space that would facilitate SDGs and biodiversity targets.

Economic Full Life-cycle Analysis is essential to conservation, as illustrated by wolves in Wisconsin. Wolves were essentially wiped out in the USA (bounties continued into the 1960s), but bounced back under protection from the US Endangered Species Act until a backlash caused their federal de-listing: 218 of 1,000 Wisconsin wolves were killed within three days. Wolves scarcely ever kill people, but they kill livestock and upset hunters. White-tailed deer numbers in America have soared (between 2011 and 2016 hunters declined by 25 per cent due to lifestyle and attitudinal shifts), leading to 1–2 million deer–vehicle collisions (DVCs) yearly

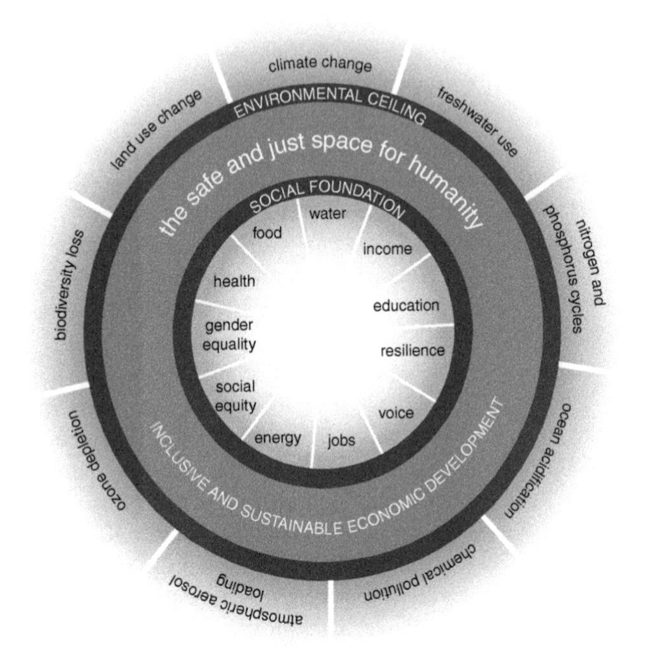

24. The doughnut of social and planetary boundaries.

(30,000 injuries, 200 deaths, and US$10 billion of costs). Wisconsin averages 19,757 DVCs, 477 injuries, and 8 human deaths p.a. From 1988 to 2010, comparing DVCs before and after wolves returned, associated wolf presence in Wisconsin was linked with a 25 per cent reduction in DVCs (two fewer human deaths), saving $10.9 million p.a., 63 times the livestock compensation of $174,000. Each of Wisconsin's 1,000 wolves annually eats 18–20 of the state's 1.6 million deer, a 1 per cent loss to the hunters. The 25 per cent cut in DVCs may reflect a 'landscape of fear': wolves travel roads (they dodge cars so suffer only 20 collisions yearly) causing deer to avoid roads. The detail of this research matters less than the lateral thinking it illustrates.

This sort of full system analysis is at the heart of natural capital thinking, incisively expounded in Dieter Helm's *Green and Prosperous Land*. Capital can be man-made, human, and natural. Ecosystems are natural capital assets and provide multiple services: they maintain a genetic library, preserve and regenerate soils, fix nitrogen and carbon, recycle nutrients, control floods, filter pollutants, pollinate crops, and operate the hydrological cycles. These vital forms of natural capital have been depleted and the situation is getting rapidly worse. The consequences are frequently hard to reverse and often non-linear, risking abrupt collapse. Furthermore, property rights over natural capital are often unclear—I was once involved in a study of potential REDD+ sites in the Indonesian island of Sulawesi which was thwarted because so many jurisdictions thought they held title over the same bit of forest.

The value of natural capital is so vast that it transforms conservation, and societal, thinking. Sometimes that value is obvious: in 2021 coral reef-related tourism in Mexico's Quintana Roo was US$3,494 million. Sometimes it requires understanding: vultures were originally considered valueless, then valued for disease control, and now for tourism. And sometimes willingness to pay a little is cumulatively worth a lot: the value of enjoying urban songbirds is estimated for Berlin and Seattle, respectively, at US$70–120 million p.a.

Natural capital includes biodiversity, of which some elements (certainly most vertebrate groups) have intrinsic value. This does not matter from a viewpoint that treats biodiversity solely as a utility to be used to benefit people. However, a non-anthropocentric economist would ask how economic decision-making should take account of both human well-being and whatever counts as fair use of non-humans (inter-species distributive justice; Figure 25). To clarify this distinction between anthropogenic and non-anthropogenic economics, in the former only humans

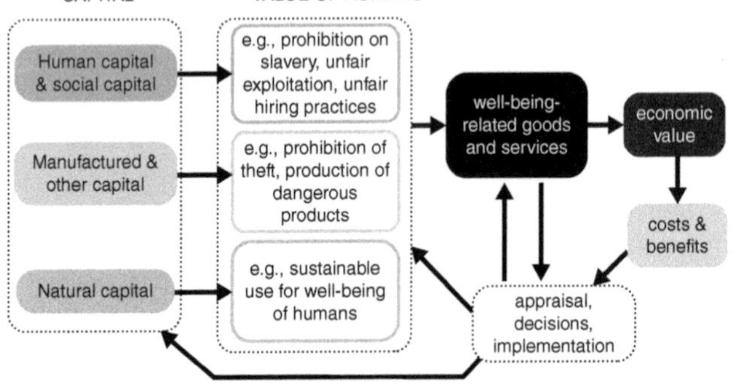

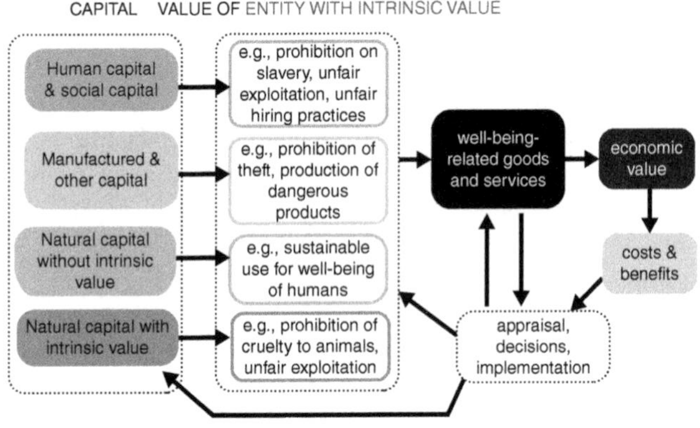

25. (a) Under anthropocentric economics the intended uses of (and impacts on) natural capital are constrained to preclude overexploitation that diminishes human well-being, and uses of human capital are constrained to preclude unfair or undignified treatment of other humans. (b) Minimally non-anthropocentric economics acknowledges natural capital with intrinsic value (e.g. most vertebrates) and constrains its uses to preclude its unfair or undignified treatment.

have chips in the banker's game, whereas in the latter other species are considered as if they do too.

From the perspectives of both intergenerational welfare and ethics, policy interventions are urgent to slow or reverse natural degradation. For over a century the main instrument to regulate biodiversity conservation has been protected areas. But mounting population growth, economic expansion, and attention to historical injustice add a focus on conservation beyond protected areas. PAs cannot be expanded sufficiently to deliver the protection of 30 per cent by 2030 (IUCN Motion 101)—this requires working with citizens within the framework of economic instruments such as taxes, subsidies, and tradable permits that can control pollution or regulate the extraction of natural resources. A specific economic instrument, payments for environmental services (PES), is increasingly relevant to biodiversity conservation.

As Johan du Toit argued in 2022, agricultural development and wildlife communities are usually incompatible due to the costs of conservation. He proposes 'a proactive shift in attitude towards the livestock–wildlife interface, from problem control to asset management'. Africa has experienced unprecedented development growth, but often at the expense of biodiversity (perversely threatening millions of livelihoods dependent on nature). In southern Kenya, where wildlife and traditional pastoralism still coexist, land price around the Nairobi National Park and Maasai Mara National Reserve is increasing the costs of maintaining and acquiring conservation buffer zones, corridors, and dispersal areas outside the protected area. This drives the transfer of land ownership from traditional pastoralists to speculators. So much so that it may be economically untenable to prevent these ecosystems from collapsing. A remote sensing analysis of over 40,000 km of extensive fencing revealed it was most prolific where community tenure had been converted to private title, especially where land values are high. This resulted in the transformation of biodiverse rangeland into agricultural, industrial, and urban uses. Planning incentives, policies, and economic forces are driving unsustainable and potentially irreversible social-ecological transitions. The continued sustainable coexistence of people and nature in African savannahs will require a range of financial, policy, and governance-related interventions.

Aligning global and local values

A fundamental challenge to biodiversity conservation is protecting species that are highly valued globally but have little, or negative, value locally. This applies particularly to megafauna which, like elephants trampling a maize crop or carnivores killing stock, cause the greatest loss to the poorest people. These charismatic species have high 'existence value' in the developed world, and generate high market value globally in terms of photographic tourism, trophy hunting, and zoo admissions. But this is rarely reflected at the local level, where communities coexisting with these animals can suffer substantial diverse costs.

In an innovative application of happiness economics, Kim Jacobsen led a team surveying people coexisting with lions in Zimbabwe. Intangible factors (such as fear and ecocentric values) were as important as, if not more important than, tangible factors (such as livestock losses) for understanding attitudes. Furthermore, socio-economic variables that are often the focus of HWC research and seem important when studied in isolation were inconsequential when statistical techniques revealed the influence of variables related to beliefs, perceptions, and past experiences. Not surprisingly, experiencing livestock loss to lions had a negative impact on happiness, and the economic value of this harm was equivalent to US$5,800 annually per person. This amount is far beyond the capacity of most development programmes to compensate communities financially. Furthermore, fear of lions had an impact on well-being, equivalent in magnitude to the effect of losing livestock to lions.

Coexistence with large carnivores can also entail significant indirect and opportunity costs. These include time and money devoted to livestock herding, guarding, and predator control. The cascading consequences may not be obvious from the comfort of a welfare state in the global north: time required for livestock protection can eat into that available for attending school, while the money lost to depredation might otherwise have paid for school fees, worsening an intergenerational transmission of poverty. The greatest toll can be in human lives—while the most people are killed by crocodiles, hippos, and elephants (typically 1,000, 500, and 100 deaths respectively p.a.), in Tanzania, where average per capita income averages around $500, 563 people were killed by lions between 1990 and 2005.

Although charismatic megafauna, and indeed birds, can generate considerable revenue, in developing countries an embarrassing flow of income streams is diverted to wealthy countries and local elites, de-incentivizing conservation on the ground. This poor cost–benefit ratio at a local level has led to the killing of metaphorical geese that might have laid a golden egg: dozens of elephants are poisoned each year in palm oil plantations in Indonesia; South American ranchers and farmers kill jaguars which they see as a threat to their livestock. The urgency of remedying this double whammy—lost wildlife and lost revenue—is emphasized by much of the remaining range of threatened large carnivores being on human-dominated land. More than 80 per cent of what remains of habitat occupied by tigers is outside reserves; more than 90 per cent for jaguars and snow leopards. The ideal is to translate the high global value of large carnivores into incentives for their conservation at a local scale, while enabling local people to escape from existing poverty traps. Too often different arms of government focus on development and biodiversity, and where they coincide the juxtaposition 'development-led conservation' makes plain which is the big brother. The challenge lies in innovating conservation-led development. A globally respected vehicle is the Darwin Initiative, announced by the UK government at the Rio Earth Summit in 1992, which supports developing countries to conserve biodiversity and reduce poverty by providing grants for projects to help them meet their objectives under the CBD and CITES.

Working across over 1 million hectares of threatened forests in 13 Chiefdoms, the Luangwa Community Forests Project and the Lower Zambezi REDD+ Project connect five National Parks. This creates a critical wildlife corridor, while transforming the livelihoods of 225,000 people across 37,000 households through revenue from forests carbon fees under the UN's REDD+ model. Within this framework the Lion Carbon project addresses the twin threats of biodiversity loss and climate change. It achieves this by adding wildlife conservation activities to sustainable management of the forests, and the creation of a premium REDD+ forest carbon credit: Lion Carbon. Lion Carbon incentivizes biodiversity conservation, providing land management planning and anti-poaching support, alongside REDD+ avoided deforestation measures, and drives the development of alternative livelihoods. Sales of Lion Carbon credits support HWC mitigation and community benefits linked to the verified presence of large carnivores and their prey: by 2022 there was a significant increase in lion, leopard, and elephant occupancy, and the project was so successful that all available credits had been sold. Indeed, we are all consumers, and the 2022 audit of Oxford's biodiversity impacts revealed that the greatest impact was through resource use and waste in upstream supply chains—this is probably true for many individuals too, and leaves the only practical option as heavy, and truly additional, biodiversity offsetting.

Financial mechanisms to realign global and local economic values

It is a classic 'market failure' that a globally valued resource is depleted because of a lack of sufficient economic incentives to maintain it locally. A solution, 'payments to encourage coexistence' (PECs), involves three steps: the species' presence is ascribed high external value, which is translated into local payments for those negatively affected by its presence to encourage human–wildlife coexistence. Two questions determine whether a PEC can benefit both people and wildlife: is there a threat to wildlife which is likely to be mitigated by the PEC, and can PEC be enacted at a scale likely to secure conservation of the target population? Three important approaches have been developed, particularly for carnivores: compensation and insurance schemes, revenue-sharing initiatives, and conservation payments.

(i) Compensation and insurance schemes. Having validated the often-exaggerated economic costs of coexistence, some agency—perhaps the international community—can offset incurred costs by paying direct compensation to individuals affected. The hope is that this reduces hardship, along with animosity toward, and retaliatory killing of, the problematic species (e.g. suspected livestock depredation by tigers is independently investigated and, if confirmed, direct compensation is paid to the farmer). My first experience of compensation was in the 1970s when the Italian government compensated shepherds for sheep killed by the handful of wolves then surviving in the Abruzzo Mountains. This memorably revealed key weaknesses: in addition to the minefield of quantifying the loss, people can harvest the compensation and can be ingenious in devising fraudulent claims. In the Italian case, our team, led by Luigi Boitani—nowadays a major force for restoring European Carnivores—had to examine dead sheep to verify whether they had been killed by wolves, or were poor quality stock on which the shepherds had set their own dogs in order to make a false claim.

Even a few hostile people can radically decrease the survival chances of a particular threatened carnivore population. Therefore, placating a particularly irate stockman may be disproportionately influential. So too may be policing the unscrupulous—in a compensation scheme for lion depredation in

Makgadikadi, Botswana, we found that a few recalcitrant individuals with poison and traps could undo all the work of a conservation programme. Compensation can also create perverse incentives, decreasing motivation to protect stock, even to the extent of increasing losses and exacerbating conflict. Lowered costs of depredation can cause people to raise their stocking rates, intensifying grazing, leading to a decline of wild prey, heightening carnivore conflict with people. When not paid, against a feeling of entitlement that may have been created by the compensation scheme itself, it can provoke anger, distrust, and ultimately retaliation.

Good governance and management of livestock must be linked to compensation. Alongside Nepal's national compensation scheme, to reduce human–felid conflicts and alleviate poverty the Living with Tigers (LWT) project engaged with eight buffer-zone communities of Nepal's Bardia and Chitwan National Parks according to the integrated, inclusive conceptual model of Figure 26.

Poorer villagers with larger families and less education ventured into the forest to gather fuel and roofing material and to graze stock, risking encountering a tiger or leopard: 12.5 per cent of livestock-owning households experienced livestock predation. LWT sought to promote coexistence by reducing livestock predation and obviating the need to enter forests. Over 4,000 people benefited and a comparison revealed that in communities where the team intervened livestock predation decreased, and predator-proof pens increased, as did mean household income by 25 per cent, along with awareness, positive attitudes, and tolerance towards felids.

Insurance schemes require participants to pay a premium (ideally reducing dependence on external funding). As with compensation, claimants often resentfully feel payments inadequately cover the

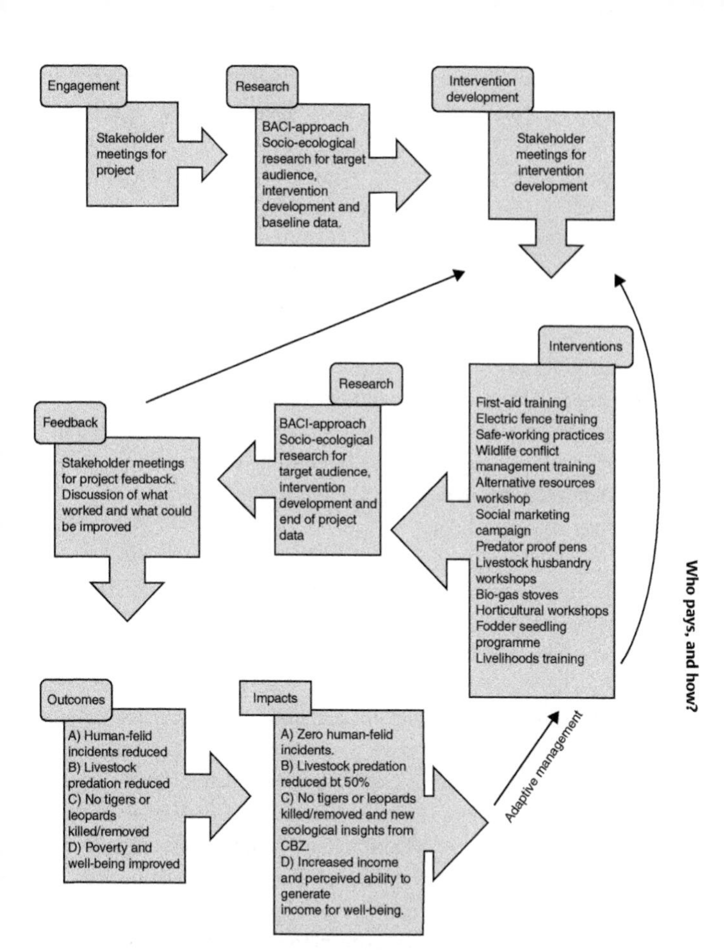

26. 'Living with Tigers' Project conceptualized pathway for interventions, intended outcomes, and impacts. Black arrows indicate adaptive management feedback loops (BACI = Before-After-Control-Intervention; CBZ = Community Buffer Zone).

direct or ancillary costs of coexistence (including direct and opportunity costs of guarding, unhappiness, and fear).

Outcomes are generally better where households are more dependent on their natural resources and livestock: a compensation scheme on Mbirikani Group Ranch in Kenya was linked to fewer lions being killed and, in India, a communal insurance and incentive scheme safeguarded snow leopards and their prey. However, in Wisconsin USA there was no difference in tolerance of wolves between those who were, or were not, compensated for their depredations. Conservationists fondly imagine that the more people know about a wild species, the more they will come to appreciate it—but despite being highly knowledgeable about wolves, ranchers around Yellowstone generally detested them.

(ii) Revenue-sharing initiatives. Revenue generated by wildlife (e.g. through photo- or hunting-tourism) can offset a community's costs of coexistence, if it reaches them. Lion numbers have increased on Namibian communal conservancies where local stakeholders retain all revenue from wildlife use. Similarly, nearly three-quarters of local people not surprisingly felt more positive about three Ugandan national parks due to tourism revenue-sharing that invested over $80,000 in schools, clinics, and infrastructure. Money is too often captured before it reaches the people who bear the costs. Around Makgadigadi Pans in Botswana the costs of lion depredation were borne by small cattle posts near the park boundary, whereas government invested the revenue from tourism in far-off hospitals. Cattlemen were unmotivated to improve stock care, and instead indicated a preference to kill lions.

Improving the cost–benefit of living adjacent to a protected area can encourage migration there, increasing competition for grazing land, settlement, and land conversion in wildlife-rich areas, and reinforcing poverty traps and degrading biodiversity. Protected areas, the 19th- and 20th-century bulwarks of conservation, may provide opportunities for nearby communities, but may also restrict their land-use options into what has been called—with obvious venom—'forced primitivism'. Households adjacent to Madagascar's Mantadia National Park endured losses of $419, half the annual per capita income, because of restricted access to agricultural land.

Although wildlife-related revenue-sharing may soften the jaws of local poverty traps, there is little evidence that it offers escape from chronic poverty or much enhances biodiversity (except where photo-tourism is most profitable: in the Maasai Mara conservancies over US$7 million p.a.—more than 50 per cent of households' annual income—reaches households directly from tourism).

(iii) Conservation payments. Compensation or revenue-sharing do not necessarily deliver measurable biodiversity conservation benefits, whereas Payments for Ecosystem Services (PES) are linked to desired outcomes. Mexican ranchers were paid between $50 and $300 if camera-traps recorded a jaguar, puma, ocelot, or bobcat on their land. Sometimes a proxy is used: in Nepal, payments are contingent upon villagers not killing snow leopards and they are rewarded for improving the habitat and abundance of bharal, an important prey. In 1996, the Swedish government initiated PES to obtain and maintain stable populations of wolves, lynx, and wolverines, mostly in Sami pastoralist rangelands; the number of certified wolverine litters in the reindeer area has now exceeded the target of 90 per year. However, it seems that culturally no payment is enough to persuade Swedish farmers to tolerate any wolves. Paying for improvements against a known baseline is easy, whereas paying incentives for conservation is difficult because additionality—the improvement against what would have happened otherwise—is hard to measure. Mechanisms that attract bidders who may seem competitively cheap but do not deliver additionality is called 'adverse selection' in economics.

An ideal PEC would minimize conflict by specifically targeting payments to those most directly affected by wildlife; reduce the direct costs of human–wildlife coexistence; provide local people with additional revenue directly linked to wildlife; avoid moral hazard and perverse incentives; not require significant additional external revenue; specifically link payments to desired conservation outcomes; and be likely to have a positive impact on human poverty. The utopia of coexistence is a situation in which local people receive tangible, commensurate, and equitably distributed benefits from wildlife that outweigh all the diverse costs.

British farmers received £700 million Pillar 2 payments in 2018 under the European Union Common Agricultural Policy to deliver environmental benefits. Following Brexit, the government sought ingenious mechanisms to do this better, brilliantly illustrated by the reverse auction to conserve turtle doves whose numbers had fallen by 98 per cent since 1970, and a further third thereafter. The auction, similar to ones used to sell radio spectrum to telecoms, allowed farmers to bid publicly for contracts to provide up to two essential resources (seeds for food, drinking water) while committing not to cut shrubby nesting hedgerows. The doves need all three, dispersed in high-quality patches in the smallest area, but that might be best provided by more than one farmer to maximize an index of Turtle Dove Happiness (abbreviated to 'Tah-Dah'). The scheme was trialled with £720,000 of government PES, attracting 63 bidders in two counties. The economic concept of the smallest area to juxtapose the best combination of necessary resources is pleasingly reminiscent of an ecological concept, the Resource Dispersion Hypothesis I first posited in 1983, which describes how animals solve exactly the same problem when configuring their territories.

Conservation pay-outs may be generated by the stakeholder, rather than coming from external sources. Predators are said to cost South Africa's farmers US$22–171 million annually, so they routinely kill jackals, caracals, and leopards, and a gruesome by-catch ranging from bat-eared foxes to aardwolves and even cheetah. A profit-and-loss account of coexisting with predators in South Africa's Eastern Cape evaluated how effective was all this killing, in comparison to non-lethal alternatives. A before–after comparison on 11 farms of a year of lethal control followed by two years of non-lethal control (guard dogs, guard alpacas, protective collars) revealed that by the second year losses of both livestock and money had been cut radically, both in terms of stock losses and running costs of the stock protection operation (Figure 27).

At least on the time-scale of the experiment, farm costs were reduced by 74 per cent by swapping to non-lethal methods. Of course, somebody still had to pay the bill: should the residual losses simply be tolerated, as part of the natural baseline of farming (much like bad weather), and could greater effort (at a cost) reduce the residual problem? A full audit might include comparison of population impact, and suffering, of the two methodologies on each species of wildlife, and the cultural value to the farmers of hunting leopards as their fathers and grandfathers have done before them. It is for wider society to decide which of these currencies has legitimacy nowadays.

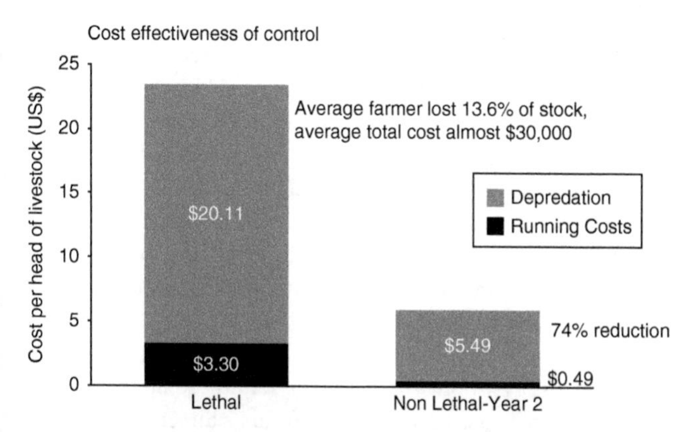

27. Cost per head of livestock incurred by attempted protection against predators by lethal and non-lethal means.

Smaller-scale economics

Funding models. The bow-hunting of Cecil in 2015 catalysed interest in lion conservation, but what would it cost, and how could it be paid for? Mindful of the umbrella potential of big cats, in 2018 a group of us estimated that to protect 1.2 million km^2 of African Protected Areas occupied by lions would cost about 1 billion US dollars a year—not a lot, in the contexts of development budgets. One mechanism that might add to government and philanthropic funds could be revenue from the symbolic use of charismatic animals in affluent economies. National animals, sports mascots, and fashion are three prominent areas of lucrative animal symbolism which illustrate the sums that could be created from a species royalty.

According to the British Egg Industry Council, 34 million eggs are consumed in Britain each day, with the British Lion Quality Seal stamped on 85 per cent of them. If each lion stamp was translated into 0.1 pence for lion conservation, this would generate £10.5 million a year. England's Premier League uses a crowned lion logo as the hallmark of its merchandise; one pound donated to lion conservation for every football shirt sold would raise about £5 million a year. As for fashion, might leopard print in the high street reflect an affinity for spotted cats that could pay for conservation? Sadly, thus far there is scant correlation between interest in leopard print and concern, or any thought at all, for leopards in the wild, but if that link could be made the potential would be huge: figures relating to leopard or animal print fabrics are difficult to come by, but as the worldwide fashion industry is worth *3 trillion* dollars annually, a minuscule royalty would transform conservation.

Small changes in ecosystem services. The value of some biodiversity can be partially calculated from its direct use, for example in the Scottish Highlands, the profit-and-loss accounts of harvesting red deer or watching beavers. The National Survey of Fishing, Hunting and Wildlife-Associated Recreation in the USA estimated that in 1996 35 million people fished, 14 million hunted, 9.5 million hunted and fished, and 63 million participated in some form of wildlife viewing: a total expenditure of US$101 billion, ~1.4 per cent of the national economy. Non-use benefits are less easily monetized, but are what matter to the conservation of most plants and animals (one whimsical proposal is to tax binoculars and bird-watching gear much like guns and ammunition, to broaden the constituency that pays for wildlife). The provision of an ecosystem service, such as conserving biodiversity, is rarely linearly related to the area of land committed to it—is 1 hectare of wetland devoted to conservation half as valuable as 2 hectares? The marginal benefit of an extra unit of habitat for an animal or plant has two parts: a direct increase in habitat quantity and an increase in probability of population persistence on the enlarged area of land. Heeding the crucial concept of marginal value, including a measure of impact on population viability accommodates the ecological reality that a small increase in habitat quantity or quality can have high value if it greatly enhances the probability of population persistence. A crucial requirement for water vole conservation is vegetated river banks and emergent vegetation: whilst the average value in 2020 of a metre of water vole habitat was £12, the marginal value of a metre could range from £0 to £40 or more, depending on its contribution to population viability.

I have characterized biodiversity conservation as spanning a continuum from groundedness to geopolitics. And it is infused with the challenge of scaling-up solutions: from riverside water vole habitat to national strategy for the oceans (which contribute US$3 trillion annually to global GDP, but are the most underfunded United Nations' Sustainable Development Goal). A historic jump in scale was made in November 2021 when The Nature Conservancy (TNC) announced an innovative US$364 million financial transaction with the government of Belize.

This enabled the country to reduce its debt and generate US$180 million for marine conservation, in return for protecting 30 per cent of its ocean, strengthening governance of domestic and high sea fisheries, and establishing a regulatory framework for coastal blue carbon projects (Figure 28). Crédit Suisse acted as the structuring bank and arranger of the financing, while the US International Development Finance Corporation provides political risk insurance to allow the transaction to achieve an investment grade rating. The loan enables Belize to repurchase and retire existing external commercial debt, create significant annual cash flows for conservation through to 2040, and establish an endowment to fund conservation thereafter. The TNC hopes to use similar debt transactions to ensure the new protection of up to 4 million km² of biodiversity-critical ocean habitats.

Ed Barbier, in his 2022 book *Economics for a Fragile Planet*, asks, 'what policies are required to "decouple" wealth creation and economic prosperity from environmental degradation, to sustain per capita welfare and simultaneously limit environmental risks?' He continues, 'we must end the under-pricing of nature so that our institutions, incentives and innovations reflect the growing ecological and natural resource scarcity'. It would be difficult to summarize the conundrum of 'who pays and how' more succinctly.

COMMITMENT
Government commits ~30% of ocean area protection, TNC restructures debt

PLANNING
Stakeholder driven planning to design protecting ocean areas

CONSERVATION
Conservation trust fund supports improved management, a healthier ocean, sustainable economic opportunities

28. Flow diagram of The Nature Conservancy (TNC) 'Blue Bonds for Ocean Conservation' programme.

Chapter 10
What next?

For a generation or more, biodiversity conservationists have been preoccupied with holding the line, fighting on all the fronts introduced in Chapters 4–9, to stop, or at least slow, the loss of biodiversity. A modern characterization of their efforts might follow the drumbeat of a three-phrase chant: stop the loss, reduce the cost, unlock the value. Nowadays, politicians, business, and a wide public understand better the importance of biodiversity within the framework of sustainability. The urgency is even more acute and a new era of biodiversity conservation is committed to greater ambition, to reversing the trend and reclaiming—sometimes literally—lost ground. An inspirational example is the jaguars of Iberá. Large carnivores are vital engines of key ecological processes, so the reduction due to habitat loss and hunting of the jaguars of Argentina to scarcely 200 individuals in 5 per cent of their historical range is tragic. The last jaguars of the Iberá wetlands, a protected area in Corrientes province, disappeared in the 1950s, along with much of their prey. A remarkable collaboration began the fight back by boosting the abundance of surviving prey—caimans, capybaras, and marsh deer—and restoring locally extinct pampas deer, giant anteaters, and collared peccaries, as part of a multi-species translocation programme. Meanwhile, the team energetically canvassed the goodwill of every sector of the community, and worked to transform the local economy from extraction based to ecotourism based. After

10 years of preparation, in 2021, they released two female jaguars, each with two 4-month-old cubs, with another 8–20 individuals queuing for release soon.

Restoration, in ecology, is the science-based assisted recovery of an ecosystem that has been degraded, damaged, or destroyed. Reintroduction involves putting a species of animal or plant back into a habitat from which it has been eliminated or, in the case of reinforcement, boosting its numbers where they are precariously low. In addition, two other forms of conservation translocation are assisted colonization and ecological replacements, which controversially involve moving species beyond their indigenous range to avert extinction or replace lost ecological functions. A currently buzzing variant is 'rewilding'—the restoration of cultivated land to a previous natural state or, in the more generic case of wilding, to a more natural state without implying return to a particular past—the intuitive meaning is perfectly captured by the subtitle of Isabella Tree's 2017 inspirational book *Wilding: The Return of Nature to a British Farm*. Rewilding was formally conceived by Michael Soulé and Reed Noss in 1998 as a mechanism for conserving biodiversity using large core protected areas, the connectivity between them, and the reintroduction of keystone species, particularly large predators. The conservation value of well-connected large protected areas harks back to the theory of island biogeography (which states that smaller, more isolated islands have fewer plant and animal species), in that large predators have special importance in driving trophic cascades and need large and connected reserves. By contrast, the return to wilderness of farmed areas through deliberate abandonment constitutes 'passive' rewilding. 'Active' variants increasingly mean conservation translocations designed to re-establish a lost or impoverished ecological process—at an extreme this can mean species' function is prioritized over form. Rewilding challenges practitioners to understand how past, present, and future human activity has, is, and will affect ecosystem function, and to take appropriate remedial action with as little perpetual intervention as possible. Because evolution plays out over geological time, a big question for theoreticians and practitioners alike is: to which desired state should they strive to restore biodiversity? Neither biology nor logic can provide a single right answer. The question becomes less about return to a lost past and more about creating an appropriate future by re-establishing naturally dynamic processes.

Just as they have suffered the worst ravages of invasive species, nowadays Australia and New Zealand are at the cutting edge of conservation translocation theory and practice, illustrated by the banded hare-wallaby and Shark Bay rufous hare-wallaby. Both species of hare-wallaby were snuffed out on the Australian mainland by non-native red foxes and feral cats, but clung on in two islands, Bernier and Dorre, of Western Australia's Shark Bay. By 2017, sheep, goats, and feral cats had been weeded out from nearby Dirk Hartog Island, facilitating a pilot trial involving 8 females and 4 males of each species. The following year, 90 banded hare-wallabies and 50 rufous hare-wallabies were transported there, in even sex ratios, and thrived. The ingredients of success were the successful removal of sheep, goats, and cats; the availability of highly suitable habitat at release sites; lessons learned from the trial translocation; the expertise of the field-team; and the simultaneous release of large numbers of translocated individuals. More experience was gained during the adaptive management of three successive trial reintroductions of endangered eastern quolls to Mulligans Flat Woodland Sanctuary, in Australian Capital Territory. The conservationists clearly learned from their experience: the survival rates in the 42 days following each trial improved from 28.6 per cent to 76.9 per cent to 87.5 per cent. Experience matters, and it is accumulating as the number of reintroductions mount across diverse taxa (Figure 29). Alongside the razzmatazz of charismatic reintroductions, 112 species of terrestrial insects have been translocated, as have over 242 oceanic species (30 per cent plants, 44 per cent coastal invertebrates). Assisted colonizations beyond indigenous range

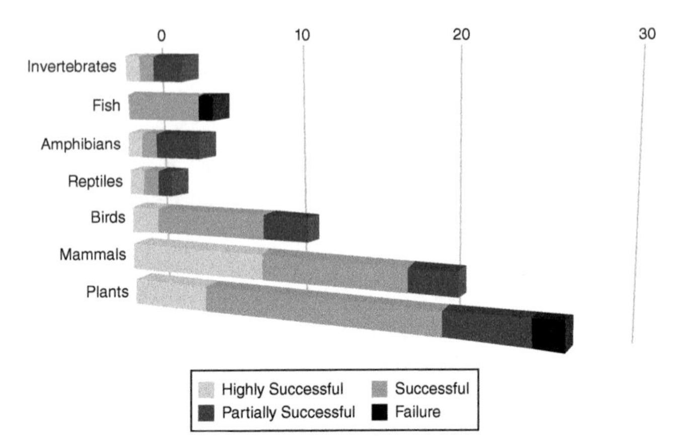

29. Success/failure of conservation translocation according to major taxa.

now outnumber reinforcements of still extant populations in pro-active responses to climate change.

What working parts might be reinserted into depleted ecosystems? My formative experience was in the Canadian prairies, working with Axel Moehrenschlager on reintroduced swift foxes—2 kg miniatures of the more familiar red fox—which painstakingly gained a foothold despite initially high death rate due to intra-guild hostility from coyotes and golden eagles. The introduction of captive-bred European mink to the Estonian island of Hiiumaa illustrates an intersection between rewilding and invasive species. Hiiumaa was populated by feral American mink, which provided a similar ecological process—mustelid predation—to that potentially deliverable by native European mink: while conservationists value the process, they value most its delivery by the right, that is the native, species. Turning from the reinsertion of a small predator to that of a large prey, in 2020 the Foundation Conservation Carpathia triumphantly reintroduced bison to Romania's Făgăraş Mountains, after an absence of more than 200 years. This followed careful public consultations, and progression through a quarantine enclosure and then a large acclimatization enclosure prior to full release. This was the first step in a plan to reintroduce a total of 100 bison in three connected areas, to seed a viable population, and thereby reconstruct the habitat's trophic chain.

As apex predators, large carnivores play crucial roles in ecosystems, yet their numbers have plummeted. Relocations could be a lifeline to restoration. Seth Thomas led an analysis of almost 300 animal relocations of 18 carnivore species spanning 22 countries in five continents, and while two-thirds had been successful, a third had not: why? Conservationists seemed to be learning: for wild-born carnivores, success rates increased from 53 per cent pre-2007 to 70 per cent; and for captive-born animals, success rates doubled from 32 per cent pre-2007 to 64 per cent. But species differed: the relocations had gone well for maned wolves, pumas, and ocelots, but badly for African lions, brown hyaenas, cheetahs, Iberian lynx, and wolves. Overall, 'soft release' (acclimatization) increased the odds of success by 2.5-fold, younger transportees did better than older, wild-born better than captive-born. Relocations will surely become a crucial arrow in the quiver of biodiversity recovery, and studies like this shine a light on how well it is working, and, importantly, how to make it work better.

The swift foxes and the European bison had completely gone from, respectively, Canada and Romania, but the situation was different for the Apennine yellow-bellied toad—an endangered Italian endemic that was disappearing due to agricultural drainage of wetland and was further blighted by chytrid fungus. These toads survived only in fragmented enclaves of 6–20 individuals. The task, therefore, was to bolster, or reinforce, their numbers as they faced all the hazards shared by small, fragmented populations. The first step, begun in 2014, was to ensure that candidate release sites were chytrid free, while meantime breeding up a captive stock raised from eggs to robust yearlings. The first batch of 67 toads was released between 2014 and 2017. Their pampered upbringing resulted in them breeding almost two years younger than their wild contemporaries. By the end of 2018 the combined offspring of the released and wild toads had doubled the original population.

Extinct-in-the-wild species' last stand

At least 77 species of plants and animals nowadays exist only in zoos and botanical collections, their futures entirely in our hands and vulnerable to demographic stochasticity (bad luck) and the genetics of inbreeding. The sihek, a kingfisher endemic to Guam, was extirpated there in 1988 by invasive brown tree snakes, and inbreeding depression damages the reproduction and survival of those in captivity. As rarity in the wild approaches hopelessness, conservationists face the dilemma of when to take the last survivors into captivity—a debate currently dividing opinion over, for example, the remaining Iranian cheetah and 42 other critically endangered species of mammalian megafauna. A review of their *in situ* extinction risk and *ex situ* management, mindful of the Convention on Biological Diversity's (CBD) Target 12 (prevent extinction of threatened species), and perils of procrastination or denial, concluded that conservationists should plan ahead for *ex situ* conservation. This is particularly important for small populations in politically unstable regions (Figure 30). When the decision was taken to catch the last few sihek, only 29 could be rescued, and dispersed amongst zoos in the United States. However, 16 of these founders had, by 2020, produced 140 descendants, albeit with a precariously imbalanced sex-ratio. Another extinct-in-the-wild Guamanian species, now back from the brink, is the flightless, endemic, Guam rail or ko'ko' bird. This bird disappeared from southern Guam in the early 1970s and from the entire island by the late 1980s. Zoo-bred ko'ko' were reintroduced to a small island near Guam, because the perils on the main island (brown snakes again) were unabated, and by 2019

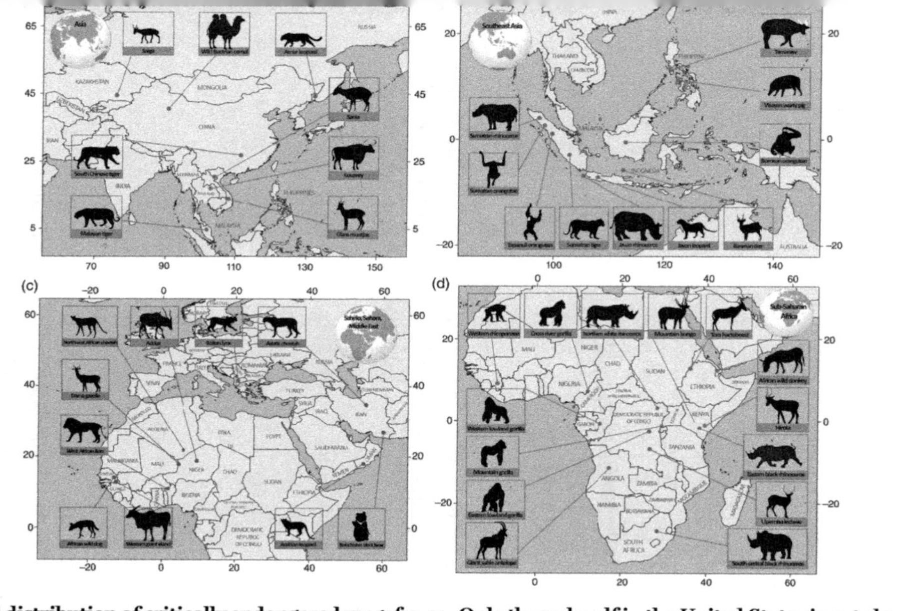

30. Global distribution of critically endangered megafauna. Only the red wolf in the United States is not shown. Almost all of these species live in hazardous circumstances, emphasizing the risks of procrastination. It's difficult to decide when to start captive-breeding, but about a third of these taxa are not yet supported by *ex situ* management, and another 23 per cent are not even supported by *in situ* programmes.

the ko'ko' had been nurtured back from extinct in the wild to critically endangered.

Rewilding Britain

About 26 per cent of Britain's land is protected, twice the proportion protected in the USA. However, this comparison is deceptive: America's protected areas include iconic national parks, such as Yellowstone, that are far closer to a naturally functioning ecosystem, with more complete trophic structures including top carnivores. Remembering the sharing/sparing dichotomy of Chapter 3, the US approach tends to land sparing, while British protected areas share the spared land. In Chapter 9, British farming was diagnosed to be addicted to subsidies that, combined with wrong-headed denial of externalities, has encouraged ruinous damage to biodiversity that should henceforth be accounted for as valuable natural capital. This realistic accountancy reveals that it is not hippyish fantasy, but good economics, for some marginal farmland to be wilded, and for all farmland to be farmed with attention to three principles: public money should be spent on public goods, natural capital should be enhanced as it is passed to the next generation, and a margin for error should be allowed—the precautionary principle. If these high-level guidelines for land-use offer a view from 40,000 feet that farming must urgently change in ways that will benefit biodiversity (and both climate and economy), how might that look at ground level?

As British conservationists have been at the forefront of advocacy for, and delivery of, biodiversity conservation around the globe, on the principle of practice what you preach, it will be appropriate for Britain to embrace enthusiastically rewilding her own backyard. However, no natural wilderness remains in Britain, and even the most glorious British countryside is man-made. The countryside Britons decide to have is a consumer choice. It may be liberating for the British to be unfettered in creating the biodiverse countryside they want, but this opportunity raises the tricky linked questions of how will the land-shared parts be managed and, for the land-spared parts, what period of the past would set the scene for the best future?

The last time the British countryside was unaffected by people, and had a full complement of megafauna, was during the last interglacial, when a mixed mosaic of open and closed vegetation communities may have prevailed. Then, during the early Holocene, 12,000 years ago, natural Britain was primarily closed woodland. But with the dawn of agriculture the balance shifted back to a mix of open wood-pasture and closed habitats, as land was cleared for agriculture and livestock were reared more extensively. After the industrial revolution, and the devastation of the post-war agricultural 'miracle', 75 per cent of Britain is now cultivated (woodland species comprise 40 per cent of those lost since 1800). Technological appreciation of restoration ecology is now sophisticated, for example under the auspices of the Conservation Translocation Specialist Group of the IUCN Species Survival Commission. The charity Rewilding Britain anticipates the imminent rewilding of 120,000 hectares.

The question of what vision of the future to draw from the past was faced by the 1,400 ha Knepp Castle Estate in West Sussex which, despite the CAP's torrent of subsidies, was losing money when farmed conventionally. Thus both finance and an ecological ethic prompted the owners' decision, in 2001, to transform it into the UK's largest lowland rewilding project. Their answer to what the future should look like was drawn from a Holocene mosaic: free-roaming horses, cattle, and pigs act, in effect, as proxies for the megafauna of 5,000 years ago—tarpan, aurochs, and wild boar. Roe deer were already at Knepp but red and fallow deer have been introduced. Each species, through their different techniques of grazing, browsing, and rootling, stimulates vegetation in different ways, creating a shifting mosaic of habitats—complexity that a single grazing species cannot generate on its own. Not only is the Estate now profitable, producing a mixture of food, education, and eco-tourism, but it has already become a nationally important breeding site for nightingales, turtle doves, and purple emperor butterflies.

Knepp illustrates ecosystem wilding at (large) farm scale, but what of a Lego block reassembly of missing elements of British biodiversity? Amongst legion opportunities, consider the field cricket which, by 1991, faced extinction in the UK. English Nature triggered a Species Recovery Programme combining habitat management where crickets survived with an *ex situ* breeding programme and health-screening protocol at the Zoological Society of London. Nymphs were taken from the wild, and between 1992 and 2007 their descendants populated a breeding programme that provided over 17,000 late-instar nymphs for reintroductions at seven sites in the cricket's former range: four were still extant up to eight generations later in 2007. In 2010 a second batch of sites each received 6 male and 6 female late-instar nymphs each April for three years, and today they are going strong.

Three biodiversity-driven ecological processes were conspicuously absent from 20th-century Britain: rootling (for creating bare ground patches for new vegetation to establish), dam building (creating niche space for other species), and predation by large carnivores (to reduce over-grazing by over-abundant prey). These services could be provided, respectively, by wild boar, by beavers, and by wolves or lynx.

Restoring ground disturbance through wild boar reintroduction

From a population estimated at about a million in the Mesolithic, the last wild boar to rootle in Britain died over 300 years ago. Then, in the late 1980s, escapees from farms bypassed the smouldering debate about whether they should be reintroduced. By 2020 there were over 4,000 wild boar at large in the UK

(1,500 in the Forest of Dean, 300 in the Weald of Kent and East Sussex, 100 on Exmoor, and 50 in west Dorset, with three populations in Scotland). For perspective, there are about 700,000 in France. Rootling is an ecological trade, whose practitioners disturb ground vegetation and top soil while foraging, akin to ploughing. Who, aside from ecologists, cares? For hunters (or restauranteurs), wild boar represent a potentially profitable quarry. Others, recalling their agricultural ravages on continental Europe or their potential as disease vectors, reflect that, in the absence of predators, boar could become a serious pest which, with a broad diet of over 400 species, could damage rare plants as well as crops. Large-scale enclosure and exclosure experiments demonstrate that wild boar could be employed as rewilding engineers, opening opportunities for less competitive plants within bracken- and heather-dominated communities. Indeed, insofar as boar support woodland regeneration, they can claim some credit for associated ecosystem services such as water regulation, carbon sequestration, and recreation (Figure 31).

Restoring river channel and riparian disturbance by reintroducing beaver

Beavers are also ecosystem engineers whose tree-felling and dam and lodge building create lakes, wetlands, and other home-made habitat for a wonderful, cascading diversity of other species. As keystone species they widen river corridors, add geomorphological complexity, biological diversity, and productivity. Sediment filters from water that seeps through the dams, creating wetlands within a mosaic landscape which, through the eyes of some beholders, is beautiful. All these good things had slipped away when, in 1191, Geraldus Cambrensis reported that the Teivi was the only river in Wales or England where beavers remained (the last Scottish beaver succumbed 400 years ago).

Stimulated by the European Habitats Directive (Article 22 required member states to restore species lost in historical times,

31. Wild boar sow with piglets: ecological engineers at work.

providing the cause of their extirpation had been removed; supported also by Article 11 of the Bern Convention), enthusiasm for reintroducing beavers started to smoulder. In 1995 I published an article entitled: 'Reintroducing the European beaver to Britain: nostalgic meddling or restoring biodiversity?' The answer rested heavily on the fact that beavers provide all three categories of ecosystem service: (a) regulatory services: improved water quality, flooding prevention, water flow regulation, raising the water table, and the conservation of water; (b) supporting services: creating and maintaining wetland habitats with benefits for biodiversity, including some economically important fish species by raising the water temperature; and (c) cultural services, such as tourism and aesthetics. Who would choose to forgo these benefits? Plenty of people, particularly those familiar with damage done by the American beaver. The beaver ponds that are beautiful and bountiful to some eyes are, to others, breeding grounds for mosquitoes and giardia, and some recalcitrant British salmon fishermen are perversely unpersuaded by their Norwegian counterparts, who generally appreciate that beaver ponds provide baby salmon (and plenty of other fish) with a perfect start in life—after all, they did evolve together. There is no more spellbinding paean to the ecosystem services provided by beaver engineering than Ben Macdonald's book, *Cornerstones*; services so advantageous that government authorities in Ohio in 1948 air-dropped beavers by parachute into remote corners of their parched state to rehydrate the landscape.

In Scotland, after a meticulously planned trial release under government licence in 2009, a beaver population is established in Knapdale Forest, as is another seeded by clandestine releases in the Forth and Tay catchments. They number about 1,000 and are a European Protected Species (ratified in Scotland in May 2019). Revealingly, human–wildlife conflict and resentment were greater where the beavers had been reintroduced clandestinely than through the careful consultations at Knapdale.

Those troubled by Malthusian nightmares of runaway population explosions might take comfort that an earlier generation of less technically or biologically educated Britons was able, seemingly easily, to over-hunt both wild boar and beavers to extinction. This suggests that both could be limited to acceptable numbers, if society was at ease with killing—indeed harvesting—them, and formed a view on what numbers were acceptable.

Restoring predation: wolf, lynx?

Wolves were once the most widely distributed non-commensal, terrestrial mammal. But in the USA their numbers went from millions in colonial times to close to zero, along with passenger pigeons and buffalo. In Europe they are recovering—there are now around 12,000 individuals across the European continent, being seen recently for the first time in over a century in Belgium, Luxembourg, Netherlands, and Denmark. The notion of reintroducing wolves to Scotland has excited British conservationists for decades. This discussion is fuelled by their reintroduction in Montana's Yellowstone National Park where elk (red deer) soon declined, and a changed landscape of fear has catalysed trophic cascades probably (there is controversy) increasing riparian woodland, along with beaver numbers, stabilizing watercourses, reducing erosion, and increasing biodiversity. In Scotland, where heavy grazing by sheep or red deer prevents regeneration of a Caledonian Forest ecosystem, introducing wolves might halve red deer abundance but would be politically unthinkable even if restructured subsidies decimate sheep numbers and launch hill farmers towards a future as skilled providers of nature. More feasibly, notwithstanding huge legislative hurdles, an extensive fenced wolf park—loosely analogous to Kruger National Park—perhaps around the famous Alladale Estate in Sutherland could be an opportunity for biodiversity, tourism, and employment (wolf reintroduction to Yellowstone was associated with a US$35.5 million increase in visitor spending). This could parallel, technically, the meta-population management of African wild dogs in South African fenced reserves. Modelling suggests that a 600 km^2 reserve might suffice, within which management might vary from no intervention (a peak of 104 wolves/1,000 km^2, suppressing deer to ~4/km^2, precipitating a crash in wolf numbers and their extinction—postponed by turning to sika and roe deer, wild boar, and beaver as available) to a limitation of only one pack/200 km^2 (an average maximum of 48 wolves/1,000 km^2, less social stresses amongst the wolves which would have little impact on deer which soar to >20/km^2). In short, it might look like wilderness, and the sound of howling wolves might make the heart sing, but even the largest imaginable fenced reserve would require intensive management with difficult operational and ethical consequences.

What of lynx? While seemingly less controversial than wolves, and successfully reintroduced in Switzerland, murmurs of lynx reintroduction to Scotland are greeted by intense opposition among those fearful for farming and conservation of endangered native species such as capercaillie and wildcat. Others rejoice at the prospect of restoring a charismatic species along with lost ecosystem processes. In polarized and acrimonious exchanges in newspapers and social media, both sides claim—without evidence—public support. To provide that evidence, conservationist Darragh Hare led an inter-sectional team to uncover public attitudes, beliefs, and policy preferences regarding lynx reintroduction amongst a stratified sample of over 1,000 people living in Scotland. Most, whether urban or rural, were moderately in favour, but nonetheless concerned about problems for livestock farmers. The debate, catapulted into politics, is a reminder that biodiversity conservation spans a spectrum from groundedness (impacts on the lives of individual organisms, and people) to geopolitics (the almost always geographically referenced interests of nations). Personally it is at the poles of that continuum, respectively with their earthiness and erudition, that I foresee the greatest scope for the future.

Across boundaries to geopolitics and the future

Biodiversity is oblivious to the often arbitrarily cookie-cut post-colonial borders of nations, but they have huge implications for its conservation. Remembering the Convention on Migratory Species (CMS), mobile birds and large mammals have horizons beyond national borders, leading conservationists into international relations and geopolitics. A population, even an individual, wide-ranging animal can fall under several political jurisdictions (with different technical expertise, knowledge, capacity, culture, and financial resources) and borderlands have dynamic social, political, economic, and sometimes even ecological transitions, which may involve armed conflict or man-made barriers. Asian leopard subspecies are all threatened, and together have transboundary populations spanning 23 countries (Figure 32). Most of the remaining fragments of Amur and Indochinese leopard ranges are in borderlands, as are about a quarter of the Persian and Arabian leopard's ranges. Their conservation hinges on transboundary collaboration. An

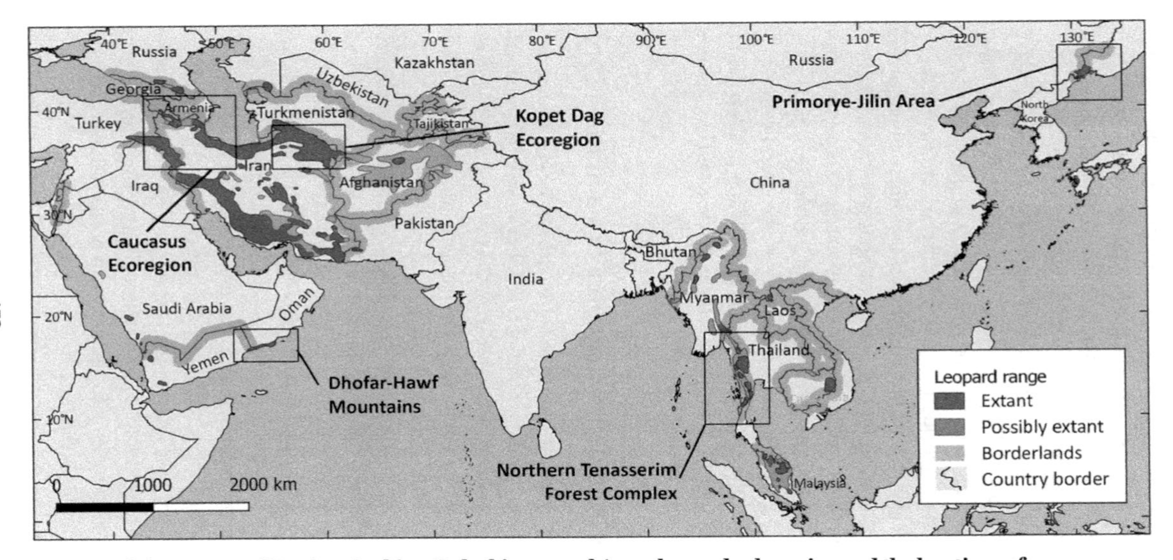

32. Remaining ranges of Persian, Arabian, Indochinese, and Amur leopard subspecies, and the locations of borderlands.

International Peace Park, linking biodiversity conservation with national security, benefits Persian leopards between Arevik National Park in Armenia and Dizmar Protected Area in Iran, and, despite political dispute between Armenia and Azerbaijan's Nakhichevan Republic, they have maintained protected areas for leopards on both sides of their border.

In everyday usage, geopolitics concerns politics, especially international relations, as influenced by geography (and thus economics and demography), often involving conflicting interests. A moment's reflection reveals this as the backcloth to, and woven through, most topics in this book, which is why I have said the new field of Conservation Geopolitics embraces almost every aspect of biodiversity conservation beyond biology. The first international conference devoted to Conservation Geopolitics was held in Oxford in 2019, with the strapline 'developing conversations across disciplines' and concerned the geographic linkages between conservation outcomes and political, social, and economic arrangements within, and relationships between, countries. Conservation Geopolitics seeks understanding of how geopolitical practices and theories affect and inform wildlife conservation. It suggests that biodiversity conservation needs a better understanding of the effects of national-level political and economic heterogeneity on biodiversity, and how this might shape geopolitical practices relating to environmental security; how geopolitical means (diplomacy, treaty-making) could improving conservation outcomes; how territories are made, borders secured, and movements of people and wildlife managed; and how to assess the efficacy of conservation strategies within existing geopolitical realities and how this might change.

Within this transdisciplinary framework, my approach to conservation is embodied in the Conservation Quartet, conceived as an antidote to Diamond's Evil Quartet. As schematized in Figure 33, partnership with local communities and stakeholders, along with education, are at the inclusive heart of this approach.

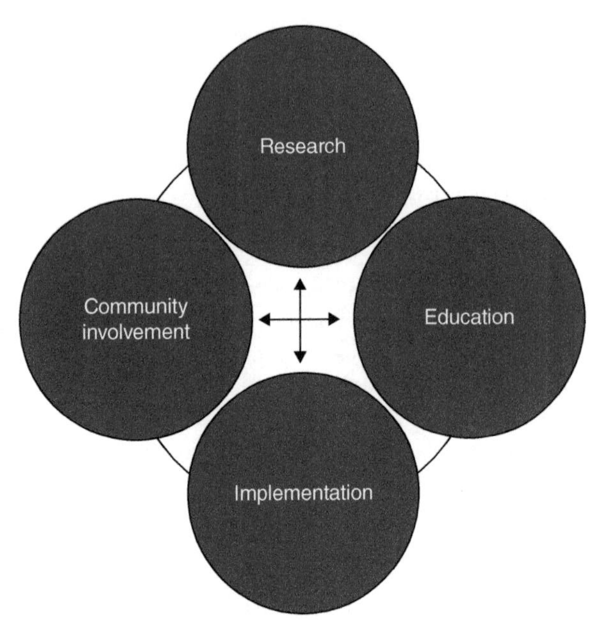

33. The Conservation Quartet, which, in 1986, I designed to conceptualize the four interacting components of the mission of the WildCRU.

The conservation quartet positions engagement of local communities in conservation as critical to its success, with inclusivity (including deliberative democracy) that is the antithesis of neo-colonialism in conservation (where the dominance of influences from outside the region unduly determine decisions within it). The importance of attending to different viewpoints was apparent in a conference I attended in June 2019 in Kigali, the Wildlife Economy Summit, to launch an 'African-led vision for conservation'. An audience including 14 Ministers of State and two Presidents, Mokgweetsi Masisi of Botswana and Emmerson Mnangagwa of Zimbabwe, heard that consumer spending on (largely wildlife) tourism in Africa of $124 billion in 2015 was set to double. President Mnangagwa's impassioned speech caught the mood and illustrated local priorities—either wildlife will pay for its future in Africa or it will go.

Opinions from the echo-chamber of shared perceptions can give a deceptive impression of unanimity. In the aftermath of the Cecil-the-lion episode, Western media and conservationists were preoccupied with divisions of opinion over the benefits of giving lions asset value versus moral repulsion at trophy hunting. In contrast, social media platforms with a predominantly African followership revealed that at least some citizens of countries where lion hunting occurred had different preoccupations. To that African constituency trophy hunting was not objectionable from an animal ethics perspective, but as a consequence of its complex historical and post-colonial associations that were interpreted as suggesting higher regard for lions than for Africans—the unintended opposite of tourism being a facilitator of intercultural understanding and peace.

There is much to celebrate in conservation, especially the internationalist spirit of talented, dedicated young conservationists, empowered by revolutionary technology (nowadays you can count elephants from space). Nonetheless, as we watch current, former, and would-be world superpowers jostling for supremacy in a world united by the internet but divided by resources and deep-seated human behaviour, towering threats loom over biodiversity, and thus over humanity. These require new approaches. Some may be resolved by Morally Contested Conservation (focusing on socially divisive issues), some by the broadening brush of holistic conservation that has long striven for the well-being of both nature and people, embracing the 'human dimension' and the double goal of economic growth and sustainability, thinking recently branded as Nature-based Solutions—defined by IUCN as 'actions to protect, sustainably manage, and restore natural or modified ecosystems, that address societal challenges effectively and adaptively, simultaneously providing human well-being and biodiversity benefits' (Figure 34).

34. Nature-based Solutions as an umbrella term for ecosystem-related approaches.

In all of these challenges and hopes it is important to understand ourselves. Humans, like all animals, have aptitudes and limitations shaped by natural selection. Knowledge of these, along with that of how mammals, including people, form societies and cooperate, can be fruitfully applied to the challenge of biodiversity conservation (and the rest of the human enterprise). Good outcomes may grow from aligning efforts with the grain of nature, rather than against it. I call this way of thinking Natural Governance and it rests on three critical pillars: (1) ecology (the dynamic balance between organisms and their environment); (2) cooperation (adaptations to act collectively even where this incurs short-term costs to self-interest); and (3) cultural systems (how these adaptations are manifested and vary across societies). Some think that human actions, and society, stem purely from political or socially constructed influences. A biological perspective does not suggest that these influences are unimportant, but asserts that people cannot be properly understood outside an evolutionary context, and nor can they deliver a future for biodiversity without understanding its processes and appreciating their oneness with nature.

Species list

aardvark *Orycteropus afer*
aardwolf *Proteles cristata*
Acacia drepanolobium
African civet *Civettictis civetta*
African elephant *Loxodonta africana*
African lion *Panthera leo*
alpaca *Vicugna pacos*
American bullfrog *Lithobates catesbeianus*
American mink *Neovison vison*
American ruddy duck *Oxyura jamaicensis*
Amur leopard *Panthera pardus orientalis*
Apennine yellow-bellied toad *Bombina pachypus*
Aphanomyces astaci (crayfish plague)
Arabian leopard *Panthera pardus nimr*
Arctic fox *Vulpes lagopus*
Asian elephant *Elephas maximus*
Asian tiger mosquito *Aedes albopictus*
Asiatic golden cat *Catopuma temminckii*
Asiatic lion *Panthera leo persica*
auroch *Bos primigenius*
banded hare-wallaby *Lagostrophus fasciatus*
basking shark *Cetorhinus maximus*
bat-eared fox *Otocyon megalotis*

bearded pig *Sus barbatus*
Bermudan snail *Poecilozonites*
bharal *Pseudois nayaur*
black-backed jackal *Canis mesomelas*
Blanford's fox *Vulpes cana*
bluefin tuna *Thunnus thynnus*
blue whale *Balaenoptera musculus*
bobcat *Lynx rufus*
brant goose *Branta bernicla*
brown bear *Ursus arctos*
brown hare *Lepus europaeus*
brown rat *Rattus norvegicus*
brown tree snake *Boiga irregularis*
buffalo *Bison bison*
Burmese python *Python bivittatus*
caiman *Caimaninae*
Caledonian crows *Corvus moneduloides*
Californian garter snake *Thamnophis sirtalis infernalis*
capybara *Hydrochoerus hydrochaeris*
caracal *Caracal caracal*
cascades frog *Rana cascadae*
cattle egret *Bubulcus ibis*
cheetah *Acinonyx jubatus*
chimpanzee *Pan troglodytes*
Chinese water deer *Hydropotes inermis*
chipmunk *Tamias*
clouded leopard *Neofelis nebulosa*
coatimundi *Nasua nasua*
collared peccary *Pecari tajacu*
coyote *Canis latrans*
crab-eating fox *Cerdocyon thous*
dingo *Canis lupus dingo*
domestic cat *Felis catus*
donkey *Equus asinus*
dotted humming frog *Chiasmocleis ventrimaculata*
dwarf mongoose *Helogale parvula*

eastern quoll *Dasyurus viverrinus*
edible dormouse *Glis glis*
elephant shrew *Macroscelididae*
Ethiopian wolf *Canis simensis*
European badger *Meles meles*
European beaver *Castor fiber*
European mink *Mustela lutreola*
fallow deer *Dama dama*
feral pigeon *Columba livia domestica*
ferret *Mustela putorius furo*
ferret badgers *Melogale*
field crickets *Gryllinae*
fig wasps *Agaonidae*
Galápagos giant tortoise *Chelonoidis niger*
Galápagos penguin *Spheniscus mendiculus*
Galápagos rice rat *Aegialomys galapagoensis*
giant anteater *Myrmecophaga tridactyla*
giant armadillo *Priodontes maximus*
giant panda *Ailuropoda melanoleuca*
giant sequoia tree *Sequoiadendron giganteum*
golden eagle *Aquila chrysaetos*
great-crested newt *Triturus cristatus*
great tit *Parus major*
great white shark *Carcharodon carcharias*
grey squirrel *Sciurus carolinensis*
Guam rail *Gallirallus owstoni*
guinea worm *Dracunculus medinensis*
Hawaiian crow *Corvus hawaiiensis*
hippo *Hippopotamus amphibius*
horseshoe bat *Rhinolophidae*
house mouse *Mus musculus*
humpback whale *Megaptera novaeangliae*
Iberian lynx *Lynx pardinus*
Indian mongoose *Herpestes javanicus*
Indian python *Python molurus*
Indian star tortoise *Python molurus*

Indochinese leopard *Panthera pardus delacouri*
Iranian cheetah *Acinonyx jubatus venaticus*
jaguar *Panthera onca*
killer whale *Orcinus orca*
king penguin *Aptenodytes patagonicus*
Kitti's hog-nosed bat *Craseonycteris thonglongyai*
large ground finch *Geospiza magnirostris*
leopard *Panthera pardus*
leopard cat *Prionailurus bengalensis*
Lord Howe Island stick insect *Dryococelus australis*
mangabey monkey *Cercocebus atys*
marbled cat *Pardofelis marmorata*
marine iguana *Amblyrhynchus cristatus*
marsh deer *Blastocerus dichotomus*
masked palm civet *Paguma larvata*
moose *Alces alces*
muntjac *Muntiacus*
musk ox *Ovibos moschatus*
nightingale *Luscinia megarhynchos*
Northern cod *Gadus morhua*
ocelot *Leopardus pardalis*
orangutan *Pongo*
osprey *Pandion haliaetus*
palm civet *Paradoxurus hermaphroditus*
pampas deer *Ozotoceros bezoarticus*
pangolins *Pholidota*
passenger pigeon *Ectopistes migratorius*
Persian leopard *Panthera pardus saxicolor*
pink pigeon *Nesoenas mayeri*
polar bear *Ursus maritimus*
polecat *Mustela putorius*
Popa langur *Trachypithecus popa*
porcupine *Erethizon*
porpoise *Phocoenidae*
prairie dog *Cynomys*
puma *Puma concolor*

purple emperor butterfly *Apatura iris*
rabbit *Oryctolagus cuniculus*
raccoon *Procyon lotor*
raccoon dog *Nyctereutes procyonoides*
red deer *Cervus elaphus*
red fox *Vulpes vulpes*
red-necked wallaby *Macropus rufogriseus*
red squirrel *Sciurus vulgaris*
red wolf *Canis rufus*
reindeer *Rangifer tarandus*
rice rat *Oryzomys*
Rice's whale *Balaenoptera ricei*
ring-necked parakeet *Psittacula krameri*
roe deer *Capreolus capreolus*
rusty patched bee *Bombus affinis*
sago palm *Cycas revoluta*
saiga *Saiga tatarica*
Santiago rice rat *Nesoryzomys swarthi*
Scottish wildcat *Felis silvestris grampia*
scrub hare *Lepus saxatilis*
sea otter *Enhydra lutris*
sea urchin *Echinoidea*
Shark Bay rufous hare-wallaby *Lagorchestes hirsutus bernieri*
ship rat *Rattus rattus*
short-clawed otter *Aonyx cinereus*
signal crayfish *Pacifastacus leniusculus*
sihek *Todiramphus cinnamominus*
sika deer *Cervus nippon*
skunk *Mephitidae*
skylark *Alauda arvensis*
small red-eyed damsel fly *Erythromma viridulum*
snow leopard *Panthera uncia*
spotted hyaena *Crocuta crocuta*
stoat *Mustela erminea*
sugar glider *Petaurus breviceps*
Svalbard reindeer *Rangifer tarandus platyrhynchus*

swift fox *Vulpes velox*
tarpan *Equus ferus ferus*
Tasmanian devil *Sarcophilus harrisii*
termite *Isoptera*
tiger *Panthera tigris*
tri-coloured bat *Perimyotis subflavus*
turtle dove *Streptopelia turtur*
Vancouver Island marmot *Marmota vancouverensis*
warbler finch *Certhidea*
warty pig *Sus cebifrons*
watermeal *Wolffia*
water vole *Arvicola amphibius*
white-clawed crayfish *Austropotamobius pallipes*
white-faced duck *Dendrocygna viduata*
white-tailed deer *Odocoileus virginianus*
wild boar *Sus scrofa*
wild dog *Lycaon pictus*
wildebeest *Connochaetes*
wolverine *Gulo gulo*
wood mouse *Apodemus sylvaticus*

References

Chapter 1: What is biodiversity, and why does it matter?

2001 Amsterdam Declaration on Earth System Science. Challenges of a Changing Earth: Global Change Open Science Conference, Amsterdam, The Netherlands, 13 July 2001.

Darwin, C., 1839. *The Voyage of the* Beagle, Natural History Library (1962 edn.). Norwell, MA: Anchor Press.

Grant, R. B. and Grant, P. R., 2003. What Darwin's finches can teach us about the evolutionary origin and regulation of biodiversity. *BioScience*, *53*(10), pp. 965–75.

Holland, P., 2011. *The Animal Kingdom: A Very Short Introduction* (Vol. 293). Oxford University Press.

Kyaw, P. P., Macdonald, D. W., Penjor, U., Htun, S., Naing, H., Burnham, D., Kaszta, Ż., and Cushman, S. A., 2021. Investigating carnivore guild structure: spatial and temporal relationships amongst threatened felids in Myanmar. *ISPRS International Journal of Geo-Information*, *10*(12), p. 808.

Macdonald, D. W., Chiaverini, L., Bothwell, H. M., Kaszta, Ż., Ash, E., Bolongon, G., Can, Ö. E., Campos-Arceiz, A., Channa, P., Clements, G. R., and Hearn, A. J., 2020. Predicting biodiversity richness in rapidly changing landscapes: climate, low human pressure or protection as salvation? *Biodiversity and Conservation*, *29*(14), pp. 4035–57.

Mora, C., Tittensor, D. P., Adl, S., Simpson, A. G., and Worm, B., 2011. How many species are there on Earth and in the ocean? *PLoS biology*, *9*(8), p.e1001127.

Myers, N., Mittermeier, R. A., Mittermeier, C. G., Da Fonseca, G. A. and Kent, J., 2000. Biodiversity hotspots for conservation priorities. *Nature*, *403*(6772), pp. 853–8.

Palmer, T. M., 2003. Spatial habitat heterogeneity influences competition and coexistence in an African acacia ant guild. *Ecology*, *84*(11), pp. 2843–55.

Chapter 2: What's the problem?

Balmford, B., Green, R. E., Onial, M., Phalan, B., and Balmford, A., 2019. How imperfect can land sparing be before land sharing is more favourable for wild species? *Journal of Applied Ecology*, *56*(1), pp. 73–84.

Barnosky, A. D., Hadly, E. A., Bascompte, J., Berlow, E. L., Brown, J. H., Fortelius, M., Getz, W. M., Harte, J., Hastings, A., Marquet, P. A., and Martinez, N. D., 2012. Approaching a state shift in Earth's biosphere. *Nature*, *486*(7401), pp. 52–8.

Bauer, H., Chapron, G., Nowell, K., Henschel, P., Funston, P., Hunter, L. T., Macdonald, D. W., and Packer, C., 2015. Lion (Panthera leo) populations are declining rapidly across Africa, except in intensively managed areas. *Proceedings of the National Academy of Sciences*, *112*(48), pp. 14894–9.

Ceballos, G., Ehrlich, P. R., and Raven, P. H., 2020. Vertebrates on the brink as indicators of biological annihilation and the sixth mass extinction. *Proceedings of the National Academy of Sciences*, *117*(24), pp. 13596–602.

Collier, P., 2008. *The Bottom Billion: Why the Poorest Countries Are Failing and What Can Be Done About It*. Oxford University Press, USA.

Diamond, J. M., 1989. The present, past and future of human-caused extinctions. *Philosophical Transactions of the Royal Society of London. B, Biological Sciences*, *325*(1228), pp. 469–77.

Dickman, A. J., Hinks, A. E., Macdonald, E. A., Burnham, D., and Macdonald, D. W., 2015. Priorities for global felid conservation. *Conservation Biology*, *29*(3), pp. 854–64.

Ecologist, The, 1972. A blueprint for survival. *The Ecologist*, *2*(1), pp. 1–43.

Erlich, P., and Walker. B., 1998. Rivets and redundancy. *BioScience*, *48*(5), p. 387.

Estes, J. A., Danner, E. M., Doak, D. F., Konar, B., Springer, A. M., Steinberg, P. D., Tinker, M. T., and Williams, T. M., 2004. Complex

trophic interactions in kelp forest ecosystems. *Bulletin of Marine Science*, *74*(3), pp. 621–38.
Horn, A. A. and Lewis, E., 1928. *The Life and Works of Alfred Aloysius Horn, an Old Visiter* (Vol. 3). London: J. Cape.
IUCN. 2021. The IUCN red list of threatened species. <https://www.iucnredlist.org/>.
Kopnina, H. and Washington, H., 2016. Discussing why population growth is still ignored or denied. *Chinese Journal of Population Resources and Environment*, *14*(2), pp. 133–43.
Leclère, D., Obersteiner, M., Barrett, M., Butchart, S. H., Chaudhary, A., De Palma, A., DeClerck, F. A., Di Marco, M., Doelman, J. C., Dürauer, M., and Freeman, R., 2020. Bending the curve of terrestrial biodiversity needs an integrated strategy. *Nature*, *585*(7826), pp. 551–6.
Lindsey, P. A., Chapron, G., Petracca, L. S., Burnham, D., Hayward, M. W., Henschel, P., Hinks, A. E., Garnett, S. T., Macdonald, D. W., Macdonald, E. A., and Ripple, W. J., 2017. Relative efforts of countries to conserve world's megafauna. *Global Ecology and Conservation*, *10*, pp. 243–52.
Macdonald, E. A., Burnham, D., Hinks, A. E., Dickman, A. J., Malhi, Y., and Macdonald, D. W., 2015. Conservation inequality and the charismatic cat: Felis felicis. *Global Ecology and Conservation*, *3*, pp. 851–66.
Poux, X. and Aubert, P. M., 2018. An agroecological Europe in 2050: multifunctional agriculture for healthy eating. *Findings from the Ten Years For Agroecology (TYFA) modelling exercise, Iddri-AScA, Study*, *9*, p. 18.
Swiss Re Group 2020. <https://reports.swissre.com/2020/>.
Wilson, E. O., 2016. *Half-earth: Our Planet's Fight for Life*. WW Norton & Company.
World Wildlife Fund. Ecoregions <http://www.worldwildlife.org/biomes>.

Chapter 3: What is the purpose of biodiversity conservation?

Arrow, K. J., Cline, W. R., Maler, K. G., Munasinghe, M., Squitieri, R., and Stiglitz, J. E., 1996. *Intertemporal Equity, Discounting, and Economic Efficiency*. Cambridge University Press, pp. 125–44.
Beston, Henry (2003 reprint). *The Outermost House: A Year of Life on the Great Beach of Cape Cod*. Macmillan, pp. 9–10.

Burke, Edmund, 1790. *Reflections on the Revolution in France*, in *The Works of the Right Honorable Edmund Burke*, vol. 3, p. 359 (1899).

Elton, C. S., 2020. *The Ecology of Invasions by Animals and Plants*. Springer Nature.

Hambler, C. and Canney, S. M., 2013. *Conservation*. Cambridge University Press.

Helm, D., 2019. *Green and Prosperous Land: A Blueprint for Rescuing the British Countryside*. HarperCollins UK.

Leopold, A., 1949. *A Sand County Almanac and Sketches Here and There*. Oxford University Press, New York.

Macdonald, D.W., 2013. From ethology to biodiversity: case studies of wildlife conservation. *Quo Vadis, Behavioural Biology*, pp. 111–156.

Macdonald, D. W., and Tattersall, F., 2001. *Britain's Mammals: The Challenge for Conservation*. People's Trust for Endangered Species.

Macdonald, D. W., Collins, N. M., and Wrangham, R., 2007. Principles, practice and priorities: the quest for 'alignment'. *Key Topics in Conservation Biology*, pp. 271–90.

Natural England, 2014. <https://assets.publishing.service.gov.uk/government/uploads/system/uploads/attachment_data/file/432726/ne-strategic-direction.pdf >.

Soulé, M. E., 1985. What is conservation biology? *BioScience*, *35*(11), pp. 727–34.

Trouwborst, A., Blackmore, A., Boitani, L., Bowman, M., Caddell, R., Chapron, G., Cliquet, A., Couzens, E., Epstein, Y., Fernández-Galiano, E. and Fleurke, F. M., 2017. International wildlife law: understanding and enhancing its role in conservation. *BioScience*, *67*(9), pp. 784–90.

Vucetich, J. A., Burnham, D., Johnson, P. J., Loveridge, A. J., Nelson, M. P., Bruskotter, J. T., and Macdonald, D. W., 2019. The value of argument analysis for understanding ethical considerations pertaining to trophy hunting and lion conservation. *Biological Conservation*, *235*, pp. 260–72.

Vucetich, J. A., Burnham, D., Macdonald, E. A., Bruskotter, J. T., Marchini, S., Zimmermann, A., and Macdonald, D. W., 2018. Just conservation: what is it and should we pursue it? *Biological Conservation*, *221*, pp. 23–33.

Vucetich, J. A., Macdonald, E. A., Burnham, D., Bruskotter, J. T., Johnson, D. D., and Macdonald, D. W., 2021. Finding purpose in the conservation of biodiversity by the commingling of science and ethics. *Animals*, *11*(3), p. 837.

Whitten, T., Holmes, D., and MacKinnon, K., 2001. Conservation biology: a displacement behavior for academia? Conservation Biology, pp. 1–3.

Chapter 4: Invasive species

Bellard, C., Cassey, P., and Blackburn, T. M., 2016. Alien species as a driver of recent extinctions. *Biology Letters*, *12*(2), p. 20150623.

Diagne, C., Leroy, B., Vaissière, A. C., Gozlan, R. E., Roiz, D., Jarić, I., Salles, J. M., Bradshaw, C. J., and Courchamp, F., 2021. High and rising economic costs of biological invasions worldwide. *Nature*, *592*(7855), pp. 571–6.

Early, R., Bradley, B. A., Dukes, J. S., Lawler, J. J., Olden, J. D., Blumenthal, D. M., Gonzalez, P., Grosholz, E. D., Ibañez, I., Miller, L. P., and Sorte, C. J., 2016. Global threats from invasive alien species in the twenty-first century and national response capacities. *Nature Communications*, *7*(1), pp. 1–9.

Faulkner, K. T., Robertson, M. P., and Wilson, J. R., 2020. Stronger regional biosecurity is essential to prevent hundreds of harmful biological invasions. *Global Change Biology*, *26*(4), pp. 2449–62.

Moorhouse, T. P., Gelling, M., and Macdonald, D. W., 2015. Water vole restoration in the Upper Thames. In Macdonald, D. W. and Feber, R. eds., *Wildlife Conservation on Farmland: Managing for Nature on Lowland Farms* (Vol. 1). Oxford University Press, USA, pp. 255–68.

Moorhouse, T. P. and Macdonald, D. W., 2015. Crayfish management in the Upper Thames. In Macdonald, D. W. and Feber, R. eds., *Wildlife Conservation on Farmland: Conflict in the Countryside* (Vol. 2). Oxford University Press, p. 165.

Chapter 5: The trade in wildlife

Bennett, E. L., Underwood, F. M., and Milner-Gulland, E. J., 2021. To trade or not to trade? Using Bayesian belief networks to assess how to manage commercial wildlife trade in a complex world. *Frontiers in Ecology and Evolution*, *9*, p. 123.

Cardoso, P., Amponsah-Mensah, K., Barreiros, J. P., Bouhuys, J., Cheung, H., Davies, A., Kumschick, S., Longhorn, S. J., Martínez-Munoz, C. A., Morcatty, T. Q., and Peters, G., 2021. Scientists' warning to humanity on illegal or unsustainable wildlife trade. *Biological Conservation*, *263*, p. 109341.

Koh, L. P., Li, Y., and Lee, J. S. H., 2021. The value of China's ban on wildlife trade and consumption. *Nature Sustainability*, *4*(1), pp. 2–4.
Macdonald, D. W., Loveridge, A. J., Dickman, A., Johnson, P. J., Jacobsen, K. S., and Du Preez, B., 2017. Lions, trophy hunting and beyond: knowledge gaps and why they matter. *Mammal Review*, *47*(4), pp. 247–53.
Macdonald, D. W., Harrington, L. A., Moorhouse, T. P., and D'Cruze, N., 2021. Trading animal lives: ten tricky issues on the road to protecting commodified wild animals. *BioScience*, *71*(8), pp. 846–60.
Macdonald, D. W., Jacobsen, K. S., Burnham, D., Johnson, P. J., and Loveridge, A. J., 2016. Cecil: a moment or a movement? Analysis of media coverage of the death of a lion, Panthera leo. *Animals*, *6*(5), p. 26.
Roe, D. and Lee, T. M., 2021. Possible negative consequences of a wildlife trade ban. *Nature Sustainability*, *4*(1), pp. 5–6.
Vucetich, J. A., Burnham, D., Johnson, P. J., Loveridge, A. J., Nelson, M. P., Bruskotter, J. T., and Macdonald, D. W., 2019. The value of argument analysis for understanding ethical considerations pertaining to trophy hunting and lion conservation. *Biological Conservation*, *235*, pp. 260–72.
World Resources Institute and International Union for Conservation of Nature, 1992. Global biodiversity strategy: guidelines for action to save, study, and use earth's biotic wealth sustainably and equitably. World Resources Inst.

Chapter 6: Wildlife disease

Anderson, R. M. and May, R. M., 1978. Regulation and stability of host–parasite population interactions: I. Regulatory processes. *The Journal of Animal Ecology*, pp. 219–47.
Bourne, F. J., Donnelly, C. A., Cox, D. R., Gettinby, G., McInerney, J. P., Morrison, W. I., and Woodroffe, R., 2007. TB policy and the ISG's findings. *The Veterinary Record*, *161*(18), p. 633.
Buttke, D., Wild, M., Monello, R., Schuurman, G., Hahn, M., and Jackson, K., 2021. Managing wildlife disease under climate change. *EcoHealth*, *18*, pp. 406–10.
Cardoso, B., García-Bocanegra, I., Acevedo, P., Cáceres, G., Alves, P. C., and Gortázar, C., 2022. Stepping up from wildlife disease

surveillance to integrated wildlife monitoring in Europe. *Research in Veterinary Science, 144*, pp. 149–56.

Macdonald, D. W. and Laurenson, M. K., 2006. Infectious disease: inextricable linkages between human and ecosystem health. *Biological Conservation, 131*(2), p. 143.

Macdonald, D. W., 1980. *Rabies and Wildlife: A Biologist's Perspective.* Oxford University Press.

Macdonald, D. W., Riordan, P., and Mathews, F., 2006. Biological hurdles to the control of TB in cattle: a test of two hypotheses concerning wildlife to explain the failure of control. *Biological Conservation, 131*(2), pp. 268–86.

Sillero-Zubiri, C., Marino, J., Gordon, C. H., Bedin, E., Hussein, A., Regassa, F., Banyard, A., and Fooks, A. R., 2016. Feasibility and efficacy of oral rabies vaccine SAG2 in endangered Ethiopian wolves. *Vaccine, 34*(40), pp. 4792–8.

Steck, F., Wandeler, A., Bichsel, P., Capt, S., Häfliger, U., and Schneider, L., 1982. Oral immunization of foxes against rabies laboratory and field studies. *Comparative Immunology, Microbiology and Infectious Diseases, 5*(1–3), pp. 165–71.

Worobey, M., Levy, J. I., Serrano, L. M. M., Crits-Christoph, A., Pekar, J. E., Goldstein, S. A., Rasmussen, A. L., Kraemer, M. U., Newman, C., Koopmans, M. P., and Suchard, M. A., 2022. The Huanan market was the epicenter of SARS-CoV-2 emergence. Zenodo. <https://doi.org/10.5281/zenodo.6299600>.

Chapter 7: Human–wildlife conflict

Arbieu, U., Mehring, M., Bunnefeld, N., Kaczensky, P., Reinhardt, I., Ansorge, H., Böhning-Gaese, K., Glikman, J. A., Kluth, G., Nowak, C., and Müller, T., 2019. Attitudes towards returning wolves (Canis lupus) in Germany: exposure, information sources and trust matter. *Biological Conservation, 234*, pp. 202–10.

Dickman, A., Marchini, S., and Manfredo, M., 2013. The human dimension in addressing conflict with large carnivores. *Key Topics in Conservation Biology, 2*(1), pp. 110–26.

Fletcher, R. and Toncheva, S., 2021. The political economy of human–wildlife conflict and coexistence. *Biological Conservation, 260*, p. 109216.

Frank, B., Glikman, J. A., and Marchini, S., eds., 2019. *Human–Wildlife Interactions: Turning Conflict into Coexistence* (Vol. 23). Cambridge University Press.

IUCN. https://www.hwctf.org/about

Macdonald, D. W. and Sillero-Zubiri, C., 2004. Conservation: from theory to practice, without bluster. In Macdonald, D. W. and Sillero-Zubiri, C., *The Biology and Conservation of Wild Canids*. Oxford University Press, pp. 353–72.

Madden, F. and McQuinn, B., 2014. Conservation's blind spot: the case for conflict transformation in wildlife conservation. *Biological Conservation, 178*, pp. 97–106.

Marchini, S., Ferraz, K., Zimmermann, A., Guimarães-Luiz, T., Morato, R., Correa, P., and Macdonald, D., 2019. Planning for coexistence in a complex human-dominated world. In *Human–Wildlife Interactions: Turning Conflict into Coexistence*. Cambridge University Press, pp. 414–38.

Nyhus, P. J., 2016. Human–wildlife conflict and coexistence. *Annual Review of Environment and Resources, 41*, pp. 143–71.

Sibanda, L., Van der Meer, E., Hughes, C., Macdonald, E. A., Hunt, J. E., Parry, R. H., Dlodlo, B., Macdonald, D. W., and Loveridge, A. J., 2020. Exploring perceptions of subsistence farmers in northwestern Zimbabwe towards the African lion (Panthera leo) in the context of local conservation actions. *African Journal of Wildlife Research, 50*(1), pp. 102–18.

Spencer, H., 1868. Social Statistics, Or The Conditions Essential to Human Happiness Specified, and the First of Them Developed. Williams and Norgate.

Su, K., Zhang, H., Lin, L., Hou, Y., and Wen, Y., 2022. Bibliometric analysis of human–wildlife conflict: from conflict to coexistence. *Ecological Informatics, 68*, p. 101531.

World Wildlife Fund. https://wwf.panda.org/discover/our_focus/wildlife_practice/problems/human_animal_conflict.

Zimmermann, A., Johnson, P., de Barros, A. E., Inskip, C., Amit, R., Soto, E. C., Lopez-Gonzalez, C. A., Sillero-Zubiri, C., de Paula, R., Marchini, S., and Soto-Shoender, J., 2021. Every case is different: cautionary insights about generalisations in human–wildlife conflict from a range-wide study of people and jaguars. *Biological Conservation, 260*, p. 109185.

Zimmermann, A., McQuinn, B., and Macdonald, D. W., 2020. Levels of conflict over wildlife: understanding and addressing the right problem. *Conservation Science and Practice, 2*(10), p. 259.

Chapter 8: Climate change

Bevanger, K. and Lindström, E. R., 1995, January. Distributional history of the European badger Meles meles in Scandinavia during the 20th century. In *Annales Zoologici Fennici* (pp. 5–9). Finnish Zoological and Botanical Publishing Board.

Campbell, R. D., Newman, C., Macdonald, D. W. and Rosell, F., 2013. Proximate weather patterns and spring green-up phenology effect Eurasian beaver (Castor fiber) body mass and reproductive success: the implications of climate change and topography. *Global Change Biology, 19*(4), pp. 1311–24.

Charmantier, A., McCleery, R. H., Cole, L. R., Perrins, C., Kruuk, L. E., and Sheldon, B. C., 2008. Adaptive phenotypic plasticity in response to climate change in a wild bird population. *Science, 320*(5877), pp. 800–3.

De Barros, A. E., Macdonald, E. A., Matsumoto, M. H., Paula, R. C., Nijhawan, S., Malhi, Y., and Macdonald, D. W., 2014. Identification of areas in Brazil that optimize conservation of forest carbon, jaguars, and biodiversity. *Conservation Biology, 28*(2), pp. 580–93.

Diamond, J. M., 1989. Overview of recent extinctions. In Western, D. and Pearl, M. C., eds., *Conservation for the Twenty-First Century*. Oxford University Press, pp. 37–41.

Fischer, E. M. and Knutti, R., 2015. Anthropogenic contribution to global occurrence of heavy-precipitation and high-temperature extremes. *Nature Climate Change, 5*(6), pp. 560–4.

Gardner, C. J. and Bullock, J. M., 2021. In the climate emergency, conservation must become survival ecology. *Frontiers in Conservation Science*, p. 84.

Hersteinsson, P. and Macdonald, D. W., 1992. Interspecific competition and the geographical distribution of red and arctic foxes Vulpes vulpes and Alopex lagopus. *Oikos*, pp. 505–15.

Macdonald, D. W., Newman, C., Buesching, C. D., and Nouvellet, P., 2010. Are badgers 'under the weather'? Direct and indirect impacts of climate variation on European badger (Meles meles) population dynamics. *Global Change Biology, 16*(11), pp. 2913–22.

Malhi, Y., Lander, T., le Roux, E., Stevens, N., Macias-Fauria, M., Wedding, L., Girardin, C., Kristensen, J. Å., Sandom, C. J., Evans, T. D., and Svenning, J. C., 2022. The role of large wild animals in climate change mitigation and adaptation. *Current Biology, 32*(4), pp. R181–R196.

Rockström, J., Beringer, T., Hole, D., Griscom, B., Mascia, M. B., Folke, C., and Creutzig, F., 2021. Opinion: We need biosphere stewardship that protects carbon sinks and builds resilience. *Proceedings of the National Academy of Sciences, 118*(38).

Shum, C. K. and Kuo, C. Y., 2010. Observation and geophysical causes of present-day sea-level rise. In *Climate Change and Food Security in South Asia*. Springer, pp. 85–104.

Trisos, C. H., Merow, C., and Pigot, A. L., 2020. The projected timing of abrupt ecological disruption from climate change. *Nature, 580*(7804), pp. 496–501.

Vargas, F. H., Lacy, R. C., Johnson, P. J., Steinfurth, A., Crawford, R. J., Boersma, P. D., and Macdonald, D. W., 2007. Modelling the effect of El Niño on the persistence of small populations: the Galápagos penguin as a case study. *Biological Conservation, 137*(1), pp. 138–48.

Wang, B., Luo, X., Yang, Y. M., Sun, W., Cane, M. A., Cai, W., Yeh, S. W., and Liu, J., 2019. Historical change of El Niño properties sheds light on future changes of extreme El Niño. *Proceedings of the National Academy of Sciences, 116*(45), pp. 22512–17.

World Wildlife Fund. https://www.worldwildlife.org/threats/effects-of-climate-change#:~:text=Sea%20levels%20are%20rising%20and,risk%20from%20the%20changing%20climat

Chapter 9: Who pays, and how?

Barbier, E., 2022. *Economics for a Fragile Planet*. Cambridge University Press.

Barrett, K., Valentim, J., and Turner, B. L., 2013. Ecosystem services from converted land: the importance of tree cover in Amazonian pastures. *Urban Ecosystems, 16*(3), pp. 573–91.

Dasgupta, P., 2021. The Economics of Biodiversity: The Dasgupta Review. HM Treasury.

Dickman, A. J., Macdonald, E. A., and Macdonald, D. W., 2011. A review of financial instruments to pay for predator conservation and encourage human–carnivore coexistence. *Proceedings of the National Academy of Sciences, 108*(34), pp. 13937–44.

Dutton, A., Edwards-Jones, G., and Macdonald, D. W., 2010. Estimating the value of non-use benefits from small changes in the provision of ecosystem services. *Conservation Biology, 24*(6), pp. 1479–87.

Good, C., Burnham, D., and Macdonald, D. W., 2017. A cultural conscience for conservation. *Animals*, *7*(7), p. 52.

Helm, D., 2019. *Green and Prosperous Land: A Blueprint for Rescuing the British Countryside*. HarperCollins UK.

Jacobsen, K. S., Dickman, A. J., Macdonald, D. W., Mourato, S., Johnson, P., Sibanda, L., and Loveridge, A., 2021. The importance of tangible and intangible factors in human–carnivore coexistence. *Conservation Biology*, *35*(4), pp. 1233–44.

Lindsey, P. A., Chapron, G., Petracca, L. S., Burnham, D., Hayward, M. W., Henschel, P., Hinks, A. E., Garnett, S. T., Macdonald, D. W., Macdonald, E. A., and Ripple, W. J., 2017. Relative efforts of countries to conserve world's megafauna. *Global Ecology and Conservation*, *10*, pp. 243–52.

Lindsey, P. A., Miller, J. R., Petracca, L. S., Coad, L., Dickman, A. J., Fitzgerald, K. H., Flyman, M. V., Funston, P. J., Henschel, P., Kasiki, S., and Knights, K., 2018. More than $1 billion needed annually to secure Africa's protected areas with lions. *Proceedings of the National Academy of Sciences*, *115*(45), pp. E10788–E10796.

McManus, J. S., Dickman, A. J., Gaynor, D., Smuts, B. H., and Macdonald, D. W., 2015. Dead or alive? Comparing costs and benefits of lethal and non-lethal human–wildlife conflict mitigation on livestock farms. *Oryx*, *49*(4), pp. 687–95.

Raworth, K., 2017. *Doughnut Economics: Seven Ways to Think Like a 21st-Century Economist*. Chelsea Green Publishing.

Scheren, P., Tyrrell, P., Brehony, P., Allan, J. R., Thorn, J. P., Chinho, T., Katerere, Y., Ushie, V., and Worden, J. S., 2021. Defining pathways towards African ecological futures. *Sustainability*, *13*(16), p. 8894.

Taylor, I., Bull, J. W., Ashton, B., Biggs, E., Clark, M., Gray, N., Grub, H. M. J., Stewart, C. and E. J. Milner-Gulland, 2023. Nature-positive goals for an organisation's food consumption. *Nature Food* https://doi.org/10.1038/s43016-022-00660-2.

TNC, 2021. <https://www.nature.org/en-us/newsroom/blue-bonds-belize-conserve-thirty-percent-of-ocean-through-debt-conversion/>.

Tyrrell, P., Naidoo, R., Macdonald, D. W., and du Toit, J. T., 2021. New forces influencing savanna conservation: increasing land prices driven by gentrification and speculation at the landscape scale. *Frontiers in Ecology and the Environment*, *19*(9), pp. 494–500.

Vucetich, J. A., Damania, R., Cushman, S. A., Macdonald, E. A., Burnham, D., Offer-Westort, T., Bruskotter, J. T., Feltz, A., Eeden,

L. V., and Macdonald, D. W., 2021. A minimally nonanthropocentric economics: what is it, is it necessary, and can it avert the biodiversity crisis? *BioScience*, *71*(8), pp. 861–73.

Chapter 10: What next?

Cohen-Shacham, E., Walters, G., Janzen, C., and Maginnis, S., 2016. Nature-based solutions to address global societal challenges. *IUCN: Gland, Switzerland*, *97*, pp. 2016–36.

Cowen, S. and Sims, C., 2021. Conservation translocation of banded and Shark Bay rufous hare-wallaby to Dirk Hartog Island, Western Australia. In Soorae, P. S., ed., *Global Conservation Translocation Perspectives, 2021: Case Studies from Around the Globe*. IUCN SSC Conservation Translocation Specialist Group, Environment Agency.

Donadio, E., Zamboni, T., and Di Martino, S., 2022. *Bringing Jaguars and their Prey Base Back to the Iberá Wetlands, Argentina* (submitted).

Farhadinia, M. S., Johnson, P. J., Zimmermann, A., McGowan, P. J., Meijaard, E., Stanley-Price, M., and Macdonald, D. W., 2020. Ex situ management as insurance against extinction of mammalian megafauna in an uncertain world. *Conservation Biology*, *34*(4), pp. 988–96.

Farhadinia, M. S., Rostro-García, S., Feng, L., Kamler, J. F., Spalton, A., Shevtsova, E., Khorozyan, I., Al-Duais, M., Ge, J., and Macdonald, D. W., 2021. Big cats in borderlands: challenges and implications for transboundary conservation of Asian leopards. *Oryx*, *55*(3), pp. 452–60.

Macdonald, D. W., Tattersall, F. H., Brown, E. D., and Balharry, D., 1995. Reintroducing the European beaver to Britain: nostalgic meddling or restoring biodiversity? *Mammal Review*, *25*(4), pp. 161–200.

Macdonald, D. W., Johnson, D. D. P., and Whitehouse, H., 2019. Towards a more natural governance of earth's biodiversity and resources. *Conservation & Society*, *17*(1), pp. 108–13.

Mkono, M., 2019. Neo-colonialism and greed: Africans' views on trophy hunting in social media. *Journal of Sustainable Tourism*, *27*(5), pp. 689–704.

Soorae, P. S., ed., 2021. *Global Conservation Translocation Perspectives, 2021: Case Studies from Around the Globe*. IUCN SSC Conservation Translocation Specialist Group, Environment Agency.

Soulé, M. and Noss, R., 1998. Rewilding and biodiversity: complementary goals for continental conservation. *Wild Earth*, *8*, pp. 18–28.

Thomas, S., van der Merwe, V., Carvalho, W. D., Adania, C. H., Černe, R., Gomerčić, T., Krofel, M., Thompson, J., McBride, R. T., Hernandez-Blanco, J., Yachmennikova, A., Macdonald, D. W. and Farhadinia, M., 2023. Evaluating the performance of conservation translocations in large carnivores across the world. *Biological Conservation*, https://doi.org/10.1016/

Tree, I., 2018. *Wilding: The Return of Nature to a British Farm*. Pan Macmillan.

Wilson, B. A., Evans, M. J., Batson, W. G., Banks, S. C., Gordon, I. J., Fletcher, D. B., Wimpenny, C., Newport, J., Belton, E., Rypalski, A., and Portas, T., 2020. Adapting reintroduction tactics in successive trials increases the likelihood of establishment for an endangered carnivore in a fenced sanctuary. *Plos One*, *15*(6), p. e0234455.

https://www.nationalgeographic.com/animals/article/why-beavers-were-parachuted-into-the-idaho-wilderness.

An expanded bibliography for all ten chapters can be found at www.wildcru.org/VSI_Biodiversity_Conservation/

Further reading

Some textbooks

A reliable, comprehensive textbook, perfect for undergraduates:

Hambler, C. and Canney, S. M., 2013. *Conservation*. Cambridge University Press

A classic ecological approach, built around population processes and based on numerous case studies of enduring interest:

Caughley, G. and Gunn, A., 1996. *Conservation Biology in Theory and Practice* (No. 333.9516 C3). Wiley–Blackwell.

An accomplished practitioner's view of conservation realities on the ground:

Jimenez, I., 2021. *Effective Conservation: Parks, Rewilding and Local Development*. Island Press.

I edited these two collections of essays to provide enduring, thought-provoking chapters on many of the main recurring topics in biodiversity conservation; good introductions for undergraduates, graduate students, and other advanced readers:

Macdonald, D. W. and Service, K. eds., 2006. *Key Topics in Conservation Biology*. Wiley Blackwell.

Macdonald, D. W. and Willis, K. J. eds., 2013. *Key Topics in Conservation Biology 2*. John Wiley & Sons.

Some case studies

Lion Hearted places human–wildlife conflict and trophy hunting in the context of lion ecology, while *Humans and Lions* provides a remarkable historical perspective to these highly contemporary issues:

Loveridge, A., 2018. *Lion Hearted: The Life and Death of Cecil & the Future of Africa's Iconic Cats*. Simon and Schuster.
Somerville, K., 2019. *Humans and Lions: Conflict, Conservation and Coexistence*. Routledge.
The Badgers of Wytham Woods and *Restoring the Balance* provide detail from long-term field studies, showing how these inform conservation debates such as the control of bovine TB and the reintroduction of wolves to Isle Royale:
Macdonald, D. W. and Newman, C. 2022. *The Badgers of Wytham Woods: Models for Behaviour, Ecology and Evolution*. Oxford University Press.
Vucetich, J. A., 2021. *Restoring the Balance: What Wolves Tell Us about Our Relationship with Nature*. JHU Press.

Human–wildlife conflict and coexistence

A comprehensive edited book on modern views on conflict and coexistence:
Frank, B., Glikman, J. A., and Marchini, S., 2019. *Human–Wildlife Interactions: Turning Conflict into Coexistence*. Cambridge University Press.
A sensitive, and anguished introduction to the dilemma of killing wonderful but abundant predators to prevent them killing wonderful but imperilled prey (the sort of 'wicked problem' that characterizes much of biodiversity conservation, challenging the intellect, breaking the heart, and provoking irreconcilable views) is provided by:
Colwell, M., 2021. *Beak, Tooth and Claw*. William Collins.

Agriculture and rewilding

The two volumes of *Wildlife Conservation on Farmland* provide a diversity of chapters on different aspects of the interface of farming and wildlife, based largely on field studies around the Wytham Estate in Oxfordshire, UK. George Monbiot's *Regenesis* is aptly subtitled *Feeding the World without Devouring the Planet*. Isobel Tree's *Wilding* is an inspiring account of wilding a large estate in lowland England, whereas Pettorelli et al.'s edited book presents wide-ranging papers from an academic conference. Ben Macdonald's *Cornerstones* deploys insightful natural history in an inspirational routemap to rebuilding ecosystems:

Macdonald, B., 2022. *Cornerstones: Wild Forces that Can Change our World*. Bloomsbury.

Macdonald, D. W. and Feber, R. eds., 2015. *Wildlife Conservation on Farmland: Managing for Nature on Lowland Farms* (Vol. 1). Oxford University Press.

Macdonald, D. W. and Feber, R. eds., 2015. *Wildlife Conservation on Farmland: Conflict in the Countryside* (Vol. 2). Oxford University Press.

Monbiot, G., 2022. *Regenesis*. Allen Lane.

Tree, I., 2018. *Wilding: The Return of Nature to a British Farm*. Pan Macmillan.

Pettorelli, N., Durant, S., and du Toit, J. T., eds., 2019. *Rewilding*. Cambridge University Press.

Economics and geopolitics

Edward Barbier's textbook rethinks economics from the starting point that biodiversity has been catastrophically undervalued. Two remarkably incisive and policy relevant books by Dieter Helm on natural capital and its value in the British countryside, and on climate change:

Barbier, E., 2022. *Economics for a Fragile Planet*. Cambridge University Press.

Helm, D., 2019. *Green and Prosperous Land: A Blueprint for Rescuing the British Countryside*. HarperCollins UK.

Helm, D., 2020. *Net Zero: How We Stop Causing Climate Change*. HarperCollins UK.

A riveting insight into the circumstances of the poorest countries, providing a framework for thinking about biodiversity conservation in the Global South:

Collier, P., 2008. *The Bottom Billion: Why the Poorest Countries are Failing and What Can Be Done About It*. Oxford University Press, USA.

Ethics

A gentle introduction to non-anthropocentric ethics by a plant ecologist, introducing ideas about respect for nature:

Kimmerer, R., 2013. *Braiding Sweetgrass: Indigenous Wisdom, Scientific Knowledge and the Teachings of Plants*. Milkweed editions.

Index

For the benefit of digital users, indexed terms that span two pages (e.g., 52–53) may, on occasion, appear on only one of those pages.

A

B

C

D

E

F

G

H

I

J

K

L

M

N

O

S

T

U

V

W

X

Y

Z